UNDERSTANDING ENERGY

UNDERSTANDING ENERGY

by

MICHAEL OVERMAN

LUTTERWORTH PRESS · GUILDFORD AND LONDON

First published 1974

ISBN 0 7188 2082 7

Computer composed and printed in Great Britain by
Eyre & Spottiswoode Ltd., at Grosvenor Press, Portsmouth

Contents

CHAPTER 1

The Nature of Energy

When the leaders of the oil-producing Arab states announced, towards the end of 1973, that they would cut down the output of oil from the oil wells of the Middle East, people everywhere suddenly realized how vital is oil in everyday life. Oil is burnt in some power stations to produce electricity. It is the fuel for the diesel engines used in many factories. Oil is used in various forms, to produce heat to warm our factories, offices and homes. It is used by ships and railway locomotives. Refined oil is used the world over to power millions of motor cars and thousands of aeroplanes.

So the world depends on oil to keep it warm on the one hand, and to keep it mobile on the other. What is the secret of this phenomenon? How does oil give up heat, and where does this locked up heat come from? How does oil make the wheels of the motor car turn? What does this locked up 'motion' imply? How did it get there?

Oil is not the only fuel we use for warmth and for power. Wood has been used as a domestic fuel since earliest times and was also burnt to raise steam for the early steam pumps of the 18th century and for the first railway locomotives of a hundred years later. When the coal mines were established coal quickly replaced wood as the principal energy source of the nineteenth century. So wood, and coal too, through the medium of heat, had been used to keep men warm and to move things — first water, and later railway wagons. Once again we are faced with the same question. How does wood and coal give up heat? Where does this heat come from?

Every intelligent reader in this technological age knows something of the answers to these questions. When the world was faced with the Arab decision to reduce oil production the newspaper and television people coined a catch-phrase to sum up the situation. 'The Energy Crisis' is what they called it. Is oil, then, the same thing as energy? Is the heat we get from oil to warm our houses a form of energy? Is the power of the petrol we put in our motor cars (or, as our friends across the sea would say, the gasolene they put in their automobiles) . . . . . is that, too, a form of energy?

We know that the answer to both our questions is yes. We know, too, that when we catch a high-flying cricket ball and the crowd roars its appreciation as the batsman starts to walk back to the pavilion — we know that the sting in our cupped hands, the jolt generated through our hands and arms by the ball we have caught, are outward signs of the energy of the ball's motion which we have arrested.

We know that when we switch on the reading lamp on our work desk at home electricity will flow and the light will come on. We know that when a comparatively small 'atom' bomb was dropped on Hiroshima, Japan, on August 5, 1945, the devastation that followed was the result of an enormous release of energy in a fraction of a second.

So energy can be warmth; it can be the power behind an engine; it can be the movement of a falling cricket ball. Energy can be the electricity that flows into the electric lamp, it can be the devastation of the atom bomb. Where, then, is the common factor? What exactly is energy?

Energy and Work
Energy is hard to define. The encyclopaedia tries to tell us in a roundabout way. Energy, it says, is what is used when work is done. This is not easy to understand. What it means is that no work of any kind can be done without some energy being used.

Energy is used when a man lifts his son on to his shoulders or kicks a football into the air. In this case the energy comes, through his muscles, from the food he has eaten. Lifting the boy and kicking the ball are examples of work being done.

Energy is used when a railway engine pulls a heavy train out of a station. This energy came, in the days of steam locomotives, from the coal burnt in the boiler. Nowadays it comes from the diesel oil used as the engine's fuel or in the form of electricity supplied direct from the power station. Pulling the railway carriages against the resistance of their wheels on the track, and against the air resistance as the train speeds up, is another example of work being done.

Energy is used when dynamite is exploded in a borehole to

break the rock when a tunnel is being built. The energy in this case comes from the explosive when it is detonated. The work being done is the splitting of the rock.

Though energy is *used* in doing work, it is not *lost*. The textbooks explain this by telling us that when energy 'disappears' in one form, it always 'reappears' in another.

When you throw up a cricket ball the energy of your muscles is given to the ball when it leaves your hand. The ball rises for some time against the force of gravity, so work is still being done after the ball has left your hand. The energy now being used is the energy of the ball's motion. When the ball reaches the top of its flight it begins to fall. Work is no longer being done because falling is the opposite of rising against gravity. The energy of motion, which was used to raise the ball, has 'disappeared' by the time it stops rising. But it reappears as the ball falls. When you catch the ball it makes a noise and stings your palm by suddenly compressing the flesh and warming it. Your hand can grow quite hot if you play 'catch' with a friend and throw a cricket ball back and forth. The harder the throw the hotter the hand. When the ball is caught the energy of motion has again 'disappeared', but now it has reappeared as noise and heat.

Diesel oil is used up when a railway locomotive pulls a train. The oil is slowly but steadily burnt. As it is burnt the energy in the oil appears as heat. This heat causes pressure in the cylinders of the engine so that the piston moves. By regulating the pressure the piston is made to turn a flywheel. Though the engine grows hot and some of this heat is transferred to the water in the engine's cooling system, some of it disappears and reappears as the energy of motion. The energy of motion in the turning flywheel is passed on to the engine's driving wheels which pull the engine along the track. When the train has reached its destination the energy of motion has once again disappeared. But though it seems to have vanished, it has only been converted, once again, into other forms of energy. Some of it has reappeared as heat in the wheel bearings of the locomotive and of the carriages of the train. They grow warm, though the heat formed in them is carried steadily away by the air around them. As the train rushes along it has to push its way through the air. Here, too, the 'disappearing' energy is reappearing as heat and noise. We cannot feel the heat because the air carries it away as fast as it is formed; we can certainly hear the noise.

In the case of the explosive used to split rock when a tunnel is being built, the energy being made use of is once again being turned into the energy of heat and noise.

Energy's Many Disguises

By now it should be a little easier to understand what we mean when we define energy as the driving force behind any kind of work. We have seen that though we can use it in many ways, it is never used up. It merely changes from one form to another. We have seen that it appears in many forms. We can get it from coal and oil. In this form we call it chemical energy. It can be turned into the energy of heat. Heat energy can be converted into the energy of motion, and this can be changed into the electrical energy which we use when we plug a heater into a power socket, or turn on the light.

When an electric lamp is alight some of the electrical energy is turned back into heat. The light bulb grows hot. The rest of the energy produces the light, which is yet another form of energy. We call it radiant energy.

What happens to light after it leaves the electric lamp? You have only to lie in the sun to know the answer. Once again it ends up as heat.

There is one form of energy which we have all heard of, but which we have not yet discussed. This is nuclear energy. It is the energy which is used to produce heat in nuclear power stations. It is the same form of energy which is released when an atomic bomb, or a hydrogen bomb explodes. We shall explain this later.

In talking about the different forms of energy one thing has become quite clear. It is that man not only makes use of it in everything he does, but that he cannot live without it.

Sources of Energy (Fig. 1)

Primitive man had only his own muscles to help him convert energy into useful work. At the beginning of human history he had only two sources of energy. One was the radiant

to last us several hundred years. In the case of oil and natural gas the situation is very different. Many believe that all our reserves will be exhausted within 20 or 30 years, though many new oil fields have been found in recent years, some of them under the sea. The reduction in oil production imposed by the Arab oil-producing states in 1973 only brought the so-called energy crisis into the limelight. In fact a world shortage of fossil fuels (this is the general term which includes coal, mineral oil and natural gas) had been foreseen by planners for some time and research had been directed to the task of finding suitable alternatives.

Alternative Sources
There are a number of other sources of energy. Hydro-electric power, wind power, tidal power, geothermal power and solar power are five that have long been known.

Of these wind power and hydro-electric power are cheap for they rely on the sun's 'free' heat which motivates climatic action which in turn causes the wind and the rain (Fig. 2).

Experts have estimated that the energy of motion of the world's rivers is sufficient to meet 75 per cent of man's present energy demand and that the winds could, in theory, be harnessed to produce twice this quantity. In fact only about 20 per cent of potential hydro-electric power, and even less of wind power is actually made use of. The trouble with hydro-electricity is that many of the best sites are too far from populated areas where energy is needed to be of practical use. The energy of the winds has never been harnessed except in a very small way.

Tidal power is obtained by converting the energy of motion of the ocean's water (under the influence of the gravitational pull of the moon) into electrical energy, by using a special form of turbo-generator. Tidal energy is 'free' and could, in theory, produce half our total energy demand. But there are relatively few sites where the tides are high enough to make the investment worthwhile, for a tidal power station is expensive, and it is inefficient if the tidal rise and fall is low.

Geothermal power has a great potential. This is obtained by tapping the heat generated all the time in the core of the

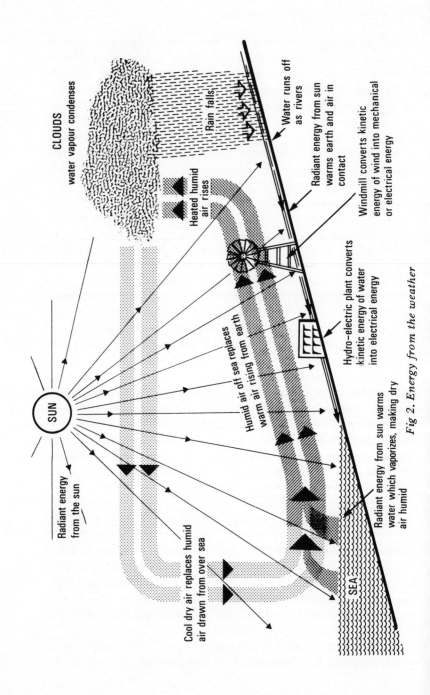

CLOUDS
water vapour condenses

Rain falls

Water runs off
as rivers

Radiant energy from sun
warms earth and air in
contact

Heated humid
air rises

Windmill converts kinetic
energy of wind into mechanical
or electrical energy

Humid air off sea replaces
warm air rising from earth

Hydro-electric plant converts
kinetic energy of water
into electrical energy

Radiant energy
from the sun

SUN

Radiant energy from sun warms
water which vaporizes, making dry
air humid

Cool dry air replaces humid
air drawn from over sea

SEA

Fig 2. Energy from the weather

earth. The world's volcanoes are visible evidence of its existence. New Zealand, Iceland, Japan and Italy already use the earth's natural heat as an energy source for the production of electricity. The geothermal power station at Lardarello converts enough earth heat into electrical energy to supply a quarter of Italy's needs. Research in this field is being conducted in California, U.S.A., and it is confidently believed that it will soon be feasible to build geothermal power stations almost anywhere in the world.

An Endless Supply

The greatest source of 'free' energy stares us, literally, in the face. It is the endless supply of radiant energy which reaches us from the sun.

The production of solar power presents technical problems of energy conversion which we shall talk about in another chapter. All that we need emphasize here is that it is virtually unlimited. All life, since the beginning of time, has depended on it. All the coal, oil and gas we burn today has been formed by conversion of the sun's radiant energy into chemical energy.

About 2 million TW (a tera watt is a million, million watts) of solar energy fall all the time on our planet. Of this about 35 per cent is reflected back into space and about 19 per cent is absorbed by the atmosphere. The remainder reaches the earth's surface, some of this being used continually to evaporate the water of the seas, lakes and rivers, and the moisture drawn out of the earth by plants and trees. Some more is converted into chemical energy by the chlorophyll in the leaves of plants and trees, the rest either turning into heat, which warms the earth, or being reflected back into space. The 'spare' solar energy totals many thousands of times the energy man can conceivably ever make use of. Research in this field is certainly being undertaken, but the fact is that at present the sun's spare radiant energy is little used by man, and there is enormous scope for engineers to devise ways of conveniently turning it into useful heat or directly into electricity.

There is one other source of energy which scientists have discovered. This is the energy stored in the atom, the nuclear energy of fission and fusion. We will explain this in a later chapter. All we need say here is that the quantity of nuclear energy available to us is in theory unlimited. Converting it to our use is a purely technical problem.

We can see now that though man's intelligence has found innumerable uses for energy, and though the world's supply of coal is by no means inexhaustible and the reserves of oil and natural gas relatively small, the energy 'crisis' which has been caused artificially by the action of some of the major oil producers is equally artificial. Unlimited supplies of energy lie all around us.

Geothermal energy, solar energy and nuclear energy are all available, if we can make use of them, in quantities immeasurably greater than our foreseeable need. If we are temporarily short this is the fault of our planners who have failed to direct research in the right directions in time.

Experience in recent years has shown that where there is a need our present day scientist-researchers, given money, can find a solution. The problem of energy supply is one of our needs and with so much 'free' energy around us, the problem of converting it into forms we can use will surely be solved.

CHAPTER 2

Energy in Disguise

Kinetic energy — the energy of motion — is easy to understand. A billiard ball rolling across the billiard table has kinetic energy, the amount depending on its weight (strictly speaking on its 'mass') and on its speed. The faster it moves the more energy it possesses. What happens when it runs into the rubber 'cushion' at the end of the table and bounces off in the opposite direction?

Potential Energy
The answer is two-fold. First the ball's motion is arrested as the rubber becomes compressed. Then the rubber returns to its normal shape pushing the ball away. What has actually happened is that the forward energy of motion has been converted into the stored energy of compressed rubber; then the latter has been reconverted into the energy of motion, which happens to be in another direction. We call the stored energy 'potential' energy. If a spiral steel spring is compressed and a strap placed around it to prevent it springing open, the stored energy will remain in the spring as long as the strap is not removed. If a cricket ball is thrown up on to a high ledge where it stays, the kinetic energy of motion has been stored as the potential energy of position. The ball could be left on the ledge for years. But if ever it were pushed off the ledge it would fall to the ground. If this happens the stored, or potential energy is released and turned once again into the kinetic energy of motion. So kinetic energy and potential energy are interchangeable. What about heat energy? Is that, too, interchangeable?

Heat
Have you ever stopped to think what heat is? What exactly is it that the burning gas gives up when you heat a saucepan of soup on the cooker? What is it that oil gives up to heat the water in the boiler of your central heating system? What is it that electricity gives up when you switch on the electric

kettle to make a cup of tea?

Many talented men had puzzled over these questions. Galileo Galilei was one of those who wanted to investigate the nature of heat. To help himself do so he made a device called the thermoscope. This consisted of a long glass tube about as thick as a straw, with a bulb the size of a hen's egg at one end. This primitive 'air thermometer' was used by heating the bulb, dipping the open end of the tube under water and then allowing the bulb to cool. As it did so water was sucked up the tube; the distance the water rose depended on the temperature to which the glass bulb had been raised above the temperature of the air. Though Galileo used his thermoscope to define what he called 'degrees of heat', he established no standard by which he could calibrate the instrument, and as air temperature and pressure varies the thermoscope only gave a rough and ready indication of temperature differences.

A French doctor, Jean Rey, tried to improve Galileo's instrument by using it the other way up and filling the bulb with water. It worked like a modern thermometer, but the tube was open at the top. The water slowly evaporated, making the thermometer inaccurate.

A further improvement was made by an Italian scientific group which called itself the Accademia del Cimento. This time the water was replaced by spirits of wine — a form of alcohol — and the long tube, formed into a spiral to make it more compact, was sealed at its upper end. The tube was calibrated with two fixed points — the temperature of ice during severe frost, and the body temperature of a healthy cow.

This Italian thermometer was certainly an improvement. But the temperature of ice can fall well below freezing point, so the calibration was arbitrary, and it was Daniel Gabriel Fahrenheit, working in Germany about 1724, who took the final step, invented the modern mercury thermometer and calibrated it using *melting* ice as his low fixed point and *man's* body temperature (which he originally fixed at 96) as the high fixed point.

The Theory of Caloric

With the Fahrenheit thermometer as a reliable new tool, the way now lay open for other men to conduct meaningful research into the nature of heat. It was a Scotsman, Joseph Black, who made several important discoveries and who was one of those responsible for the 'caloric' theory.

Black established three things. He showed that different substances had a different 'capacity for heat'. This explained why a bar of iron held in a flame grows quickly hot, while it takes much longer for the same flame to heat up a pan of water. Water's 'capacity for heat', Black correctly reasoned, is much greater than that of iron. Today we speak of the *specific heat* of a substance. Water, we say, has a specific heat of 'one'. Iron, by comparison, has a specific heat (at normal room temperature) of 0.107. This figure tells us that a fixed quantity of water needs almost ten times as much heat as the same weight of iron to increase its temperature by the same amount.

Joseph Black next established the fact that however much heat you 'put into' a saucepan of water, it stops growing hotter as soon as it starts to boil. All the heat you now put into the water is 'used up' in making the water slowly boil away. Black also showed that the same thing happened with melting ice. Water freezes and ice melts at a fixed temperature and whatever heat you put into melting ice, this does not alter its temperature. The heat is 'used up' in making the ice melt. After pondering over these last two discoveries Black wrote: 'I imagined that, during boiling, heat is absorbed by the water and enters into the composition of the vapour produced from it, in the same manner as it is absorbed by ice in melting, and enters into the composition of the produced water.' Today we call this 'absorbed' heat the 'latent heat of vaporization', and the 'latent heat of fusion'.

Two Frenchmen, quite independently, made another discovery. In 1787 Jacques Charles, a physicist who worked in Paris, showed that gases expand in proportion to the heat put into them. Joseph Louis Gay-Lussac, who had not read of Charles' work, came to the same conclusion in 1802. This seems obvious to us today. It explains why hot air rises, for it follows that when a gas expands it grows less dense. It also explains the working of the petrol engine; the heat generated when the fuel burns expands the gas in the cylinder creating a pressure which then forces the piston down.

The discovery that gases expand when heated led the scientists of the time to check whether this also happened in the case of solids and liquids. They found it was true in most cases. (Ice, as we know, provides an exception to the rule. Very cold ice does expand when heated. But when it has warmed up to $-4°C$ it then begins to contract until it is warmed up to melting point.)

The three basic discoveries led scientists, in the late 18th century, to believe that heat was a kind of fluid, and they called this fluid 'caloric'. They said that caloric exists naturally in the universe and cannot be created or destroyed. They said that caloric could be absorbed to a varying degree by other substances (this explained specific heat), and that it could be combined 'chemically' with certain substances when it changed them from a solid into a liquid form, or from a liquid into a gas (this explained latent heat). Caloric, they added, was invisible and had no measurable weight.

These eighteenth century physicists were, of course, wrong. Yet over a hundred years before, in 1620, Sir Francis Bacon had described heat as *motion*. 'Heat itself,' Bacon had written in a moment of perception, 'its essence and quiddity, is motion and nothing else.' Robert Boyle, another English scientist, was of the same opinion, and his colleague Robert Hook had stated that heat is 'nothing else but a very brisk and vehement agitation of the parts of a body'.

Bacon, Boyle and Hook were right. Today we know that heat is nothing more than the energy of motion. Mechanical energy is the energy of motion of bodies of the sizes man can naturally perceive. The cricket ball, thrown through the air, acquires kinetic energy. A satellite orbiting the earth has kinetic energy, and it continues to orbit because in space there is no air resistance to convert that energy slowly into heat. The earth itself, orbiting round the sun has kinetic energy. The tiniest grain of sand can have kinetic energy — this is the basis of its use in sand-blasting to clean up metal surfaces.

Molecules in Motion

Now consider moving particles smaller than grains of sand — molecules for example. Moving molecules certainly have kinetic energy; and it is this motion of molecules that we humans feel as heat.

To appreciate this explanation of heat let us take a closer look at moving molecules. The simplest example is found in the molecules of a gas. Suppose we have a small square box, 10 cm x 10 cm x 10 cm in size, and fill it with hydrogen. A hydrogen molecule consists of two hydrogen atoms linked together. These molecules move about rapidly all the time, their speed depending on the temperature of the gas. Whenever a molecule hits one of the walls of the container it bounces off. The effect of all the molecules continually bouncing against the six walls of the box is felt as a pressure on them. If the hydrogen is cooled the molecules travel more slowly and the pressure of the gas on the container walls is less. If the gas is heated the molecules travel faster and the pressure is greater. This is the simple explanation of the discovery made by Jacques Charles and by Louis Gay-Lussac. It explains the pressure in the cylinder of a petrol engine when the fuel and air mixture is ignited by the sparking plug and burns in a flash releasing a great deal of heat. If it seems strange that infinitesimally small molecules can acquire so much kinetic energy we must remember that the actual number of molecules involved is unimaginably large. Every cubic millimetre of hydrogen, at normal temperature and pressure, contains approximately 26,800 million molecules. Our 1,000 cm^3 box would therefore contain one million times that number.

When a piece of copper is heated the copper molecules cannot move around freely as in a gas. They are held captive in a crystal structure. Instead the kinetic energy which each heated molecule acquires appears as a rapid vibration around its fixed position in the metal. Copper conducts heat rapidly because when a copper molecule vibrates it quickly passes on the vibration to its neighbours. If you put one end of a copper bar into a flame the molecules at that end immediately begin to vibrate, and the vibration is passed quickly from one molecule to the next until all the molecules in the bar are all vibrating vigorously. If you touch the hot copper bar with your finger, the vibrations of the copper molecules disturb the skin and nerve molecules of your finger tip, making these vibrate in turn. It is this vibration in the molecules of your nerve endings which causes a message to be sent along them to your brain and which you 'feel' as heat.

Metals and other solid objects expand when heated because the increasing speed of their molecular vibration enlarges the basic dimensions of the crystal structure. When a solid is heated more and more there comes a point when the energy of molecular vibration is great enough to overcome the forces that hold the molecules in their fixed position. The result is first softening, and eventually melting of the solid.

In liquids the molecules can move about more freely for there is no longer a crystal structure to hold them rigidly in a fixed position. They bounce against each other continually and in doing so pass on part of their kinetic energy to their neighbours. When a molecule in a liquid comes up against the surface it feels much more attraction from the concentrated molecules below it than from the fewer air molecules above. Molecules reaching the surface are therefore mostly drawn back into the liquid, though some escape into the air — which explains why liquids slowly evaporate.

As the liquid is heated the speed of its molecules is increased and there comes a point when their kinetic energy is sufficient for most of those reaching the surface to escape into the air.

This is the point at which the liquid starts to 'boil'. Any more heat added now will no longer raise the temperature of the liquid. For every molecule which now has its speed increased reaches the critical temperature which enables it to escape into the air. Instead of its temperature rising further the liquid begins to boil away.

Although Daniel Fahrenheit's thermometer was properly calibrated with two fixed points — the temperature of melting ice (32°F) and of man's normal body temperature (originally taken as 96°F, but later corrected to 98.4°F) — a Swedish physicist, Anders Celsius, thought it would be simpler to use the freezing and boiling points of water as the fixed points, calling these 100° and 0° respectively. Why he

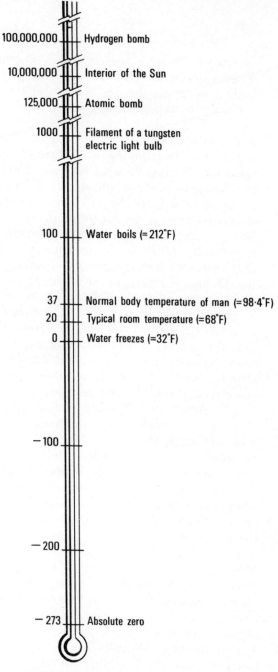

Fig 3. *The Centigrade scale of temperature*

inverted the scale is not clear, but the principle was sensible. 0° became the temperature of melting ice, and 100° that of boiling water on the new Celsius or Centigrade scale which was adopted as part of the modern metric system (Fig. 3).

The Absence of Heat

If temperature is merely a measure of the kinetic energy of molecules in motion, what happens if they grow so cold that they cease to move or vibrate at all? The answer is that their temperature is then 'absolute' zero, the temperature at which molecules possess no energy at all. Absolute zero can easily be calculated. Experiment shows that the pressure of a gas is lowered by 1/273 of its pressure at 0°C for every degree centigrade that its temperature is reduced; so at −273°C a gas should have no pressure at all, implying that all molecular motion has ceased. Absolute zero is therefore −273°C.

The domestic refrigerator is based on the principle of heat removal. Basically it is a box insulated (usually with foamed polystyrene) so that heat cannot easily get in through the walls. Ice-cold brine is circulated through the pipes in the ice chamber and as heat always flows from warm to colder objects, any heat in the refrigerator flows into the brine which emerges slightly less cold than it enters. A mechanical device is then used to cool the brine to its original temperature before it is pumped back into the refrigerator ice chamber coils. This device uses evaporation (usually of ammonia) to cool the brine. When liquid ammonia is pumped through a fine hole into an area of lower pressure it vaporizes. This vaporization absorbs heat (the latent heat of vaporization) which is supplied by the brine emerging from the refrigerator.

Storing Heat

Heat energy can be stored in small quantities by the simple expedient of feeding it into a substance and surrounding this with a heat insulating layer of some other substance, or alternatively by a vacuum.

The vibrating molecules in a thermos flask filled with hot water cannot transfer their vibrations to the air around because the space between the double glass skin of the flask contains no molecules (or very very few) to take up and pass on the vibrations.

We have seen how heat is nothing more than kinetic energy at a molecular level; we have also seen how this molecular energy produces the driving force in the cylinder of a motor car engine. By pushing the pistons down and so rotating the crankshaft of the engine the kinetic energy of heated molecules is converted into mechanical energy. The latter is seen first as the kinetic energy of the rotating flywheel, which we can compare with the vibrating molecules in a heated solid. Then, by means of gears, this rotary motion is converted into motion in one direction as the motor car runs along the road. This linear motion, too, is kinetic energy. It is comparable, in this case, to that of the molecules in a warm gas. So the conversion of heat energy into mechanical energy is, in fact, nothing more than the transferring of the kinetic energy of billions of molecules into the kinetic energy of one large object.

Electricity

Just as heat was at one time thought to be a kind of weightless fluid which scientists called caloric, electricity was believed to be a kind of 'electric fluid'. It was about 1746 that Benjamin Franklin attempted to give a scientific explanation for the phenomenon of static electricity. He suggested that when amber is rubbed with a dry woollen cloth, it becomes deficient of 'electric fluid', and that when glass is rubbed with silk it becomes 'charged' with an excess of 'electric fluid'.

We now know that though Franklin had the right idea, his theory was inverted. For it is the charged amber which has the excess, and charged glass the deficiency. Nor was 'fluid' the right word, for electricity consists, as we also now know, of free electrons — the smallest of the three basic particles of which all matter is made.

When the Italian scientist Luigi Galvani accidentally discovered, in 1786, a means of producing a continuous current of electricity, and when (fourteen years later) Alessandro Volta, also in Italy, invented the 'voltaic pile' — the world's first battery — what these two men had done was to identify a chemical method of producing a continuous supply of free electrons.

Later it was discovered that electrical energy could be converted into heat energy by passing an electric current through a conductor having a high electrical resistance. (This, of course, is the principle of the modern electric fire, of the electric cooker, of the electric smoothing iron and of many other twentieth century appliances.)

Electrons in Motion
We have seen already that heat energy is, in fact, kinetic energy — the energy of molecules in motion. What, then, is electrical energy?

The fact that a very large quantity of electrical energy can be carried along a copper wire is explained in much the same way as we explained the considerable heat energy carried by molecules. For though the electron is very much smaller than the tiny molecule, free electrons are present in conductors in very much larger numbers and can move at very much greater speeds.

A molecule of hydrogen consists of two hydrogen atoms, each made up of one proton (the nucleus) and one orbiting electron. Each atom has an overall diameter of about one fifty-millionth of a centimetre. Each proton, at the centre of each atom, has a diameter of about one ten-thousandth of that of the complete atom. And each electron weighs about one two-thousandth of the weight of the proton. So the electron is inconceivably small.

Against this the number of electrons in one centimetre of copper wire one millimetre in diameter is estimated to be in the region of 2,000 million million million, of which about 100 million million million are free to move along and so carry energy along the wire. These figures are too staggering to be readily comprehensible. To appreciate the actual capacity of free electrons to carry energy we have only to remember that a copper wire one millimetre in diameter can easily carry enough electrical energy at 240 volts to run a one horse power motor or operate a one kilowatt electric heater. For this flow of energy about 25 million million million electrons have to pass into the electric motor or heater each second.

I mentioned the 'voltage'. This is because voltage is a measure of the energy carried by electrons flowing in an electric circuit. You can compare voltage in electricity with temperature in heat; or with the pressure in a water pipe.

There is a point here that needs further explanation. If you could mark an electron and measure the speed at which it moves along an electric wire in a direct current circuit, you would find that it may take several seconds for it to travel a few centimetres. This speed has no relation to the energy which that electron carries. It is the voltage, or electrical pressure, behind it which gives each electron a kind of vibration which is 'passed' from electron to electron along a length of wire in the form of a wave. Because of the electron's vibration the negative charge carried by every electron produces a vibrating magnetic field, and this field causes the charges carried by the neighbouring electrons, and so the electrons themselves, to vibrate in sympathy. The 'wave' of vibration which travels through the wire in this manner moves at 300 million metres a second — the speed of light.

Stored Electrical Energy

Electrical energy, like the kinetic energy of motion of bodies, can be converted into stored, or potential electrical energy. A capacitor is a device which does this. A charged capacitor holds an excess of vibrating free electrons on its 'negative' side and, provided the insulation is good, they will remain there until connected to a circuit. Needless to say the capacity of the device to hold electrical energy is limited by its size. In practice capacitors are used to hold only very small amounts of stored energy, and their insulation will normally only prevent the escape of electrons carrying energy at relatively low voltages. For example transistor radio capacitors are rarely rated at above 50 volts, those used in television having insulation that will withstand up to, perhaps, 1,000 volts. In either case the capacity is so small that it can be discharged of all its energy in, literally, a fraction of a second. In nature a cloud can act as a large natural capacitor, and can store electrons at energy levels up to many millions of volts. Their capacity, too, is much higher than

CHAPTER 3

Energy in Other Guises

Though we have spoken briefly of stored energy the three forms of energy we have so far examined in detail have been examples of energy in motion. There is a fourth form of energy in motion which we all know of, but which is as different from electrical energy as this is from heat. It is *radiant* energy — the energy which travels in the form of light, or of radio waves, or ultra-violet or infra-red radiation, or of X-rays or gamma rays. All these are electromagnetic waves, their only true difference lying in their frequency.

Electromagnetic Radiation

That waves of this kind carry energy is obvious from the example of sunlight which not only lights, but warms everything on which it falls. That sunlight carries a very considerable amount of energy is seen when you focus its rays with a magnifying lens. The concentrated heat can make paper, or even wood, burst into flame. The heat produced by sunlight cannot have arrived from the sun as vibrating molecules because there are no molecules in space. Yet sunlight, and the radiant heat that comes with it, travels through millions of kilometres of space.

It was the English astronomer, Sir William Herschel, who first established, when using a prism to split a beam of sunlight into its various colours about the year 1800, that red light produced more heat where it fell than light of any other colour, and that even more heat was generated in the darkness immediately beyond the band of red produced by the prism. He reasoned, correctly, that, in addition to visible light, sunlight contains radiation having frequencies of vibration which the human eye cannot 'see', though the nerve endings in one's skin can detect the heat produced by them. This radiation was named 'infra-red' because its frequency was found to be lower than that of red light.

We saw in the previous chapter that when an electron vibrates the movement of the electric charge it carries creates a pulsating magnetic field of influence. In a conductor, where there are billions of free electrons close to each other, every vibrating electron is close enough to other free electrons for the pulsating magnetic field to exercise its influence on the charges carried by them, causing them to vibrate in sympathy. In this way electrical energy is transferred along a conductor.

In empty space a pulsating magnetic field cannot transfer its energy to electrons because there are no electrons present. James Maxwell, a brilliant Scottish mathematician, wrote a 'Treatise on Electricity and Magnetism' which he published at Aberdeen University in 1873. In this he suggested that when a magnetic field fades it transfers its energy into an electric field, and that when an electric field fades it transfers its energy into a magnetic field. This finding was in accordance with the law which states that energy can never be lost; it can only appear to vanish by being converted into another form.

Maxwell's argument implied that a pulsating magnetic field creates a pulsating electric field, and vice versa. The pulsating electric field in turn creates a pulsating magnetic field. And so the process goes on. The result is an 'electromagnetic' wave which travels through space at an enormous speed. James Maxwell showed, by means of mathematics, that the speed of this wave of radiation was the same as the speed of light. He concluded, correctly, that light was itself an example of electromagnetic radiation. It was later shown that infra-red and ultra-violet radiation, radio waves, X-rays and gamma rays consist of the same electromagnetic radiation as light, though the human eye cannot detect them, as their frequencies are outside the range which affects the nerve endings of the retina of the eye.

Once vibrating electrons have transferred their energy (or at least some of it) into electromagnetic radiation this speeds through space until it comes once again in contact with free electrons, wherever they may be. Then, just as happened in an electric conductor, the pulsating magnetic field immediately exercises its influence on the charges carried by those electrons, causing the latter to vibrate in sympathy. In this way some of the energy of the electromagnetic radiation is changed back into the energy of vibrating electrons; energy has thus been transferred by the waves (whether they be light or radio waves, or any other form of electromagnetic waves) from one lot of vibrating electrons to another lot that may be millions of kilometres away.

Electromagnetic radiation can be produced in a number of ways. The ordinary electric lamp and the gas discharge tube both give out light rays. A radio transmitter produces radio waves. High voltage electrical bombardment of heavy metals results in the emission of X-rays. Gamma rays are produced when radioactive elements break down forming new, lighter, more stable elements.

Radiant Energy in Fixed Quantities

Long before James Maxwell published his theory of electro-magnetic waves, scientists had puzzled over the nature of light. In the late 17th century Sir Isaac Newton believed that light consisted of particles. Two hundred years later, in 1901, Dr. Max Planck, a German physicist, showed that infra-red radiation (radiant heat) was given off by heated objects in tiny indivisible 'packets' of energy which he called quanta. This means that it is impossible for heat radiation to exist in quantities other than an exact number of quanta. You cannot have ½ a quantum of heat, or 10¼, or $127\frac{3}{8}$ on the other hand you can certainly have 1 quantum, or 10 quanta or 127.

When Albert Einstein published his Theory of Relativity in 1905 he not only agreed with Dr. Planck, but said that light followed a similar principle — that it could only exist in fixed quantities. Einstein called the unit of light energy the photon, the 'size' of which varies according to the frequency of the radiation. Now it can be argued that all electro-magnetic waves, including light, consist of streams of these particles, these photons, which are very much smaller than electrons, and which travel at the speed of light. This idea, strange though it may appear, provides a link between electromagnetic energy and the energy of moving bodies, or vibrating molecules and of vibrating electrons, by making it possible for us to conceive of the energy of electromagnetic waves as the energy of vibrating photons.

Chemical Energy

When a fuel is burnt heat energy is released. It is true that we
have to put in some heat to start the process of burning, but
this is only to speed up the chemistry of oxidation. Old
newspapers turn brown with age; in growing brown the paper
is becoming slowly oxidized by the oxygen present in air.
When you put a match to paper the heat accelerates the
oxidation process and the paper discolours quickly. Suddenly
it bursts into flame. This is the beginning of what we call a
chain reaction. At this point the heat of the burning match is
no longer needed because the oxidation process is itself
producing enough heat to keep the paper burning. Where
does this heat come from? Why does paper, or wood burn?
Why does oil or natural gas burn? What happens when a fuel
is oxidized?

We know already that energy cannot be created. It can
only change from one form to another. This means that fuels
must contain stored energy. So now the question is: where
does this energy come from and how is it stored?

The answer lies all around us. In Chapter 1 we spoke of the
action of sunlight on plants and trees. The leaves of plants
contain the green substance chlorophyll. This has the remarkable power of easily doing something which man can only do
with difficulty by heating water to 3,000°C. Chlorophyll
splits the water molecule (H_2O) into its constituent parts,
hydrogen and oxygen. The hydrogen joins chemically with
carbon dioxide (CO_2) present in the air forming new com-
pounds called carbohydrates. There are many different carbohydrates; they form the basis of all the common foods we
eat, including starch and sugar.

Hydrogen and oxygen form an explosive mixture. Once a
chain reaction is started by means of a flame or a spark,
oxidation of hydrogen takes place rapidly, water being
formed and a great deal of heat being released. In fact one
gram of hydrogen combines with 8 grams of oxygen to
produce 9 grams of water and 68,400 calories of heat. This is
the quantity of heat needed to raise the temperature of
68,400 grams of water by one degree centigrade — or to raise
the temperature of 68.4 grams of water from freezing point
to boiling point.

If the production of 9 grams of water from its components releases 68,400 calories of heat, it follows, since energy cannot be created or destroyed, that 68,400 calories of heat energy are needed to split 9 grams of water into hydrogen and oxygen. We have seen that the chlorophyll in plants has the ability to do this, releasing the hydrogen needed to form carbohydrates by combination with carbon dioxide. And it is sunlight that provides the energy needed for 'photosynthesis', as the process is called. This explains why plants grow much more quickly in bright sunshine than in cold cloudy weather. It also explains why they stop growing altogether at night.

The process of photosynthesis is a process of natural energy storage which is far more efficient than any method devised by man. It is, simply, the long-term storage in quantity of what we call chemical energy.

Atoms and Molecules

To understand how chemical reactions either use, or release energy, we must take a look at the nature of matter.

We are taught that all things, all substances, are made up of the basic particles of matter — protons, neutrons and electrons. The atoms of the various elements each consist of a nucleus made up of varying numbers of protons and neutrons bunched tightly together, with a varying number of electrons 'vibrating' around in orbits. While an electron orbit is normally of fixed diameter its axis varies rapidly so that the electron's path traces out the surface of a sphere. So fast does the electron travel (approximately 2,400 km/second) and so small is the orbit (the hydrogen atom's only electron normally has an orbit one ten millionth of a millimetre in diameter) that the electron moves over every part of the surface of its orbital sphere millions of times a second. We can only conceive of it as being everywhere on the surface of that sphere all the time; and we think of its path as an electron 'shell'.

Each electron, we know, carries a negative electrical charge, and each proton a positive charge. As these opposite charges attract one another there is a constant electrostatic force trying to pull the electron towards the nucleus. However, the centrifugal force caused by the high speed orbiting of the electron exactly balances the electrostatic attraction so that the electron remains vibrating around its orbital shell. (We have a man-made example of this kind of balance of forces every time a satellite is put into orbit around the earth. In the case of the satellite the force of gravity, which tries to pull the satellite back to earth, is exactly balanced by the outward centrifugal force resulting from the satellite's rapid circular motion around the earth.)

Alternative Orbits

I used the word 'normally' when referring, above, to the diameter of the orbital shell of the single electron of a hydrogen atom. In fact there is a whole series of alternative orbital shells in which the electron of a hydrogen atom can vibrate. The alternatives have a diameter four times, nine times, sixteen times and twenty five times that of the 'normal' orbit. While an orbiting electron can be 'pushed' from any one shell to any larger shell, it cannot vibrate anywhere *between* two shells. An electron vibrating in any given shell carries a fixed amount of energy, and to 'push' it into a larger shell can only be done by giving it extra energy. If it has been 'pushed' into a larger shell it has been moved against the electrostatic force which holds every electron in its normal shell. Much the same is true of the satellite, for rocket power is needed to put an orbiting satellite into a larger orbit, which means 'pushing' it further from the earth against the force of gravity.

Because an electron vibrating in a larger-than-normal shell has been pushed electrically 'up hill' it naturally tends to run electrically 'down hill' again, if it is not somehow held in its new position. So electrons in larger-than-normal orbits tend to return to their normal ones, and as they 'fall' back they give up the energy which was used in pushing them out. This energy appears in the form of electromagnetic waves. This is what happens in the ordinary electric light bulb. The filament of the electric lamp contains the metal tungsten. When electrical energy is fed into a tungsten filament some of the orbiting electrons of the atoms of this metal are pushed into larger-than-normal orbits. They immediately fall back into normal orbits, giving out light. The process is continuous.

There are various 'rules' which orbiting electrons obey. For example, while it is possible for two electrons to vibrate simultaneously in the inner, normal, shell of any atom, this is the maximum number which is ever found in this shell. Larger shells, however, can be occupied by more than two electrons. The second shell, for example, can be occupied simultaneously by any number up to eight. The third shell can hold up to eighteen. Indeed each larger shell has its own fixed maximum capacity. Another 'rule' limits the capacity of the outermost shell of any atom, however many shells it may have, to eight electrons.

An atom of the gas helium has a nucleus of two protons and two neutrons 'locked' closely together; vibrating around them are two orbiting electrons sharing a single shell. As the normal inner shell of any atom can accommodate two, and only two, electrons, this makes helium gas a 'stable' substance. This is because the chemical nature of any substance depends almost entirely on the number of electrons in its outer shell, and on whether that shell contains the maximum number possible, or less. Helium's outer shell (which is also its inner and only shell) is 'full', so the gas is stable. Hydrogen's outer shell has only one electron; so there is room for another. This makes hydrogen gas relatively unstable, which explains why it burns fiercely.

Shared Electrons

The oxygen atom is more complex. It has a nucleus consisting of eight protons and eight neutrons. The eight positive charges on the protons are balanced by the negative charges of eight orbiting electrons. Two of these electrons orbit in the inner shell, and as this is the maximum number that can be accommodated there, the remaining six orbit in the next shell. Any atom's second shell can accommodate up to eight electrons. So in the case of oxygen there is room for two more. This means that oxygen can easily combine chemically with atoms having two electrons in their outer shell (which means there is room for six more); or alternatively with pairs of atoms each having one electron in its outer shell.

If an oxygen atom links up with two hydrogen atoms, the oxygen atom can share the single electron of each hydrogen atom, giving it a total of eight electrons in its outer shell. The two shared electrons also now orbit both the hydrogen nuclei, so that the hydrogen atoms' shells are also 'full' with two electrons each. The result is a molecule of water, H_2O, which is extremely stable.

We have already seen that this process of combining hydrogen and oxygen (which is accomplished simply by 'burning' the hydrogen) produces a great deal of heat energy. When water is decomposed an equally large amount of energy has to be used to force the atoms apart from the 'embrace' of their shared electrons. It is this energy which appears to vanish, but which is in fact stored as 'chemical energy'. It is stored energy which can be taken out of store by reversing the chemical process. In the case of carbo-hydrates and the various fuels we use, this is done simply by oxidation, and oxidation can be speeded up by heating the fuel to the point where a chain reaction sets in. Heat speeds up a chemical reaction because only atoms, and not mole-cules, react together. Oxygen exists normally as molecules, each consisting of two atoms linked together by sharing outer orbit electrons. When the molecular vibration of heat energy is sufficiently violent, oxygen molecules split into individual atoms. Because these are unstable they seek other unstable atoms with which to share outer orbit electrons. The oxida-tion process itself releases heat and when this heat is sufficient to split oxygen molecules the oxidation process becomes the chain reaction we call burning.

It is not only oxidation that releases chemical energy. An atom of the metal sodium has a solitary electron in its outer orbit; an atom of chlorine has seven. Seven and one make eight — the exact number that fills any outer orbit. As a result sodium and chlorine will easily combine, giving out heat energy and forming the stable compound sodium chloride, NaCl, which is common salt.

Nuclear Energy
Just as the kinetic energy of the motion of bodies had its counterparts at molecular level (heat energy) at sub-atomic level (electrical energy) and at the sub-electronic level of the photon (radiant energy), so has chemical energy its counter-

parts. We have already spoken of the stored energy of position (potential energy) and we have now explained the stored energy at molecular level (chemical energy). Are there forms of stored energy at sub-atomic and, perhaps, at the level of the photon? The answer, as you may have guessed, is that there certainly is stored energy at sub-atomic level and, possibly, at the level of the photon too. Sub-atomic energy is what we call nuclear energy (the phrase 'atomic energy' is misleading) and it is in many ways comparable to chemical energy. For just as chemical energy is the stored energy which arises from the electrostatic force which normally keeps vibrating electrons in the orbital shells to which they belong, nuclear energy is the energy stored in the nucleus of atoms; it is the energy which normally holds protons and neutrons together in an atom's nucleus, despite the fact that all protons bear positive electric charges which repel one another.

The discovery of nuclear energy began in the year 1896 when a French physicist, Henri Becquerel, discovered 'radio-activity'. He had conducted an experiment in which he placed some uranium ore in sunlight with a photographic plate wrapped in black paper underneath it. An 'image' of the uranium was found on the plate when it was developed and he assumed that sunlight had caused the uranium to emit energy in a form which could penetrate black paper. Clouds covered the sun when Becquerel was ready to repeat the experiment and he placed the wrapped plate in a drawer with the uranium ore still lying on it. Later when he developed the plate he again found an 'image' of the uranium, though the package had remained in darkness. He had established that uranium gives off radiation which can penetrate black paper, and that it does so entirely of its own accord.

Gamma Radiation

Many scientists followed Becquerel's discovery by experimenting on similar lines. In 1897 Marie and Pierre Curie, also French (although Marie was Polish by birth), discovered two other elements, radium and polonium, which gave out similar radiation. They established that the radiation was of three kinds which they called alpha rays, beta rays and gamma rays

after the first three letters of the Greek alphabet. Alpha and
beta rays turned out to consist of electrically charged
particles, but the gamma rays were found to be electromagnetic radiation of even higher frequencies than those of
X-rays which were already known.

Subsequent research by Ernest Rutherford and his
associates at the Cavendish Laboratory at Cambridge established, in 1919, that gamma radiation was produced when the
nucleus of an atom splits, forming smaller nuclei with fewer
protons and neutrons. Rutherford's first success was in
turning nitrogen into oxygen and hydrogen.

Natural decay of radioactive elements is a relatively slow
process, just as the oxidation of newspaper at ordinary
temperatures is slow. But just as oxidation can be speeded up
to a point where a chemical chain reaction begins and the
paper burns of its own accord, so can radioactive decay be
speeded to the point where atoms begin to split spontane-
ously in a nuclear chain reaction. The result is the release of
enormous quantities of energy, partly in the form of electromagnetic radiation, but mainly as heat.

The modern nuclear reactor provides an example of the
nuclear chain reaction. The fuel used in the first nuclear
reactor, built in the U.S.A. in 1942 by a domiciled Italian
physicist, Dr. Enrico Fermi, was the heavy metal
uranium-235. The uranium-235 atom is rather complex. It
has 92 orbiting electrons and a correspondingly large nucleus
made up of 92 protons and 143 neutrons. When a
uranium-235 nucleus is hit by a free neutron travelling at
high speed (the actual speed is critical; it must not be too fast
or too slow) it is split literally in two, forming two new
smaller nuclei which share out the original atom's protons,
neutrons and electrons to form atoms of two entirely new
elements. In splitting, some of the immensely powerful
nuclear energy which normally holds protons and neutrons
together is released, in the form of intense heat and some
electromagnetic radiation (which includes dangerous gamma
rays).

In this process of splitting a uranium-235 atom, two or
three of its original neutrons are not needed to form the new
elements, and these are shot out of the splitting atom at high

speed. If these neutrons hit the nuclei of other uranium-235 atoms at the correct speed the process is repeated. Once control of neutron speed is established the chain reaction continues, producing immense heat as atoms of uranium-235 are continually split.

How the process is controlled in a nuclear reactor will be explained in a later chapter. All we need say here is that if uranium-235 is made into a sufficiently large solid lump the chain reaction, once started, can get immediately out of control. This is what happens in the atom bomb, where all the uranium-235 nuclei are split almost instantaneously, generating a colossal quantity of heat and considerable radiation. (In fact the first atom bomb used plutonium, a similar man-made heavy metal, in place of natural uranium-235.)

Nuclear energy can be released by the fusion of the nuclei of light atoms, as well as by splitting those of heavy atoms. This is what happens in the interior of the sun, and in the hydrogen bomb.

We have seen that a normal hydrogen atom consists of a single proton orbited by a single electron. Deuterium (which is chemically the same as hydrogen, though it is twice as heavy — hence its nickname, 'heavy' hydrogen) has a nucleus consisting of one proton and one neutron, orbited by one electron. Tritium is a third, even heavier form of hydrogen, the atoms of which each have in their nuclei one proton and two neutrons. Under intense heat (a temperature of about 100 million degrees centigrade is necessary) a deuterium nucleus and a tritium nucleus will fuse to form a helium nucleus, which contains two protons and two neutrons. This process, like the nuclear fission process, results in the release of an immense quantity of heat energy — so much that once started a chain reaction is easily achieved. It is this kind of chain reaction that has been going on in the sun for millions of years. (Incidentally, the fusion process also produces free neutrons, since one deuterium and one tritium nucleus together have three neutrons, though only two are needed to form a helium nucleus.)

Mass into Energy

In both fusion and fission one would expect the products to 'weigh' the same as the materials which produced them. Careful measurements have shown otherwise.

When a uranium nucleus splits, the mass of the two new smaller nuclei produced, along with the mass of the free neutrons released, proves to be slightly less than the original mass of the uranium nucleus. Somewhere in the reaction some matter appears to have vanished.

The same is true in the case of fusion. The mass of a helium nucleus, along with that of a free neutron, works out to be slightly less than the mass of a deuterium and a tritium nucleus. Once again the process seems to result in the disappearance of matter.

In fact this 'disappearance' was predicted in 1905 by Albert Einstein, whose Theory of Relativity stated, among other startling assertions, that mass and energy were interchangeable. This, indeed, is the meaning of Einstein's famous equation $E = mc^2$, in which E is a measure of energy, m is a measure of mass, and c is the speed of light.

The basic matter of which all atomic particles are made may therefore be thought of as energy stored at a sub-atomic (indeed at sub-electronic) level. Einstein proved mathematically that this was so. And because the constant 'c' (the speed of light) was so large, Einstein's equation predicts that a minute mass of basic matter is equivalent to a very great quantity of energy. The tiny loss of mass resulting from the nuclear fission and fusion processes, and the corresponding release of enormous quantities of heat, has confirmed Einstein's prediction.

Of what the particles of sub-atomic matter are made — the protons, the neutrons, the electrons and the other miscellaneous particles which physicists have identified in recent years — no one knows. Perhaps the truly basic substance of all matter is a *particle* of energy itself. Perhaps it is this particle — something similar to our vague conception of the photon — which is pure energy, stored when not in motion, but which becomes what we know as radiant energy when it travels at the speed of light. Whatever may be the truth, the idea rounds off our look around the theoretical world of energy,

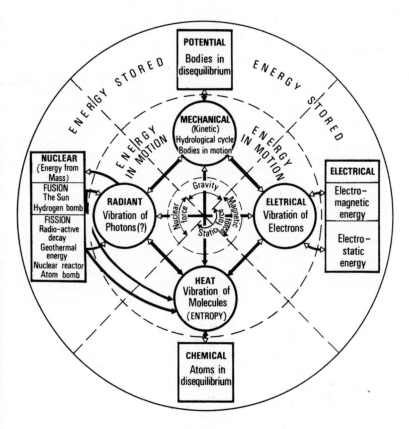

*Fig 4. The Energy Circle showing forms of energy and their
interchangeability*

and explains, except for one feature, the diagram on this page
(Fig. 4).

Entropy

The unexplained feature of the diagram lies in the black and
white headed arrows. Energy can be transferred or trans-
formed wherever arrows are shown. But there is a natural law
— the Second Law of Thermodynamics — which states, in
technical language, the rather obvious fact that heat cannot
and will not flow, of its own accord, from a colder to a
hotter substance. On the contrary 'down hill' for heat is the

opposite direction. Left to its own devices, the kinetic energy of heat, while obeying the First Law of Thermodynamics (which states that energy can never disappear — that it can only be altered into other forms) invariably dissipates itself over a wider and wider area by exciting slower vibrating molecules to vibrate a little faster. And as all energy transformations result in some production of heat this means, in effect, that all energy will one day be dissipated as heat and that all matter will eventually settle down to vibrate at the same energy level. Scientists call this natural process 'entropy'.

This fact of entropy leads us to a sobering thought: it means that a day will come when the solar system dies out, not because energy is lacking, but because it can no longer be made available where it is wanted. Not that we need be worried. As long as the sun continues to shine on us the fact of 'entropy' — natural heat dissipation — cannot affect us, for whatever heat we 'lose' will always be replaced by the sun — at least for many millions of years to come.

In the diagram (Fig. 4) the black-headed arrows indicate the natural conversion of other energy forms into heat; the white-headed arrows indicate changes which can be induced, though they do not occur of their own accord.

CHAPTER 4

Energy Sources and Resources

Today man relies on five main sources of energy. The fossil fuels — coal, oil and natural gas — account for no less than 95 per cent of world-wide consumption, the remainder coming from hydro-electric and nuclear power stations (the former providing about 3 per cent and the latter 2 per cent of the total). Oil, of which the world's reserves are unlikely to last more than 50 years, even allowing for the recent discovery of large new oil fields in Alaska and under the sea, accounts for slightly more than half of the 'free' world's entire energy consumption. (The reserves I refer to do not include the oil which is obtainable from the vast deposits of tar sands in Canada and of oil shales found mainly in the United States and Brazil. The recent 'energy crisis' has focused new attention on these deposits which have not hitherto been exploited due to the high cost of recovering oil from them.)

Figure 5 shows how dependent the advanced countries are on the fossil fuels, especially on oil. In fact, with the exception of electrified railway lines, coal burning ships and a handful of nuclear powered ships, the entire world's transport depends on oil as its energy source. Most modern ships use heavy fuel oil, most railway locomotives use diesel oil, all aircraft use light fuel oil, and virtually all motor vehicles use either light fuel or diesel oil. What will life be like, then, fifty years from now, when there is no natural mineral oil left? Perhaps oil will then be made from coal. This is possible, but expensive. If oil is extracted from the American oil shales this should multiply the world's known oil reserves by a factor of about four. But this oil would be expensive to extract, just as is oil made from coal. What alternatives can we use as fuel for transport when oil runs out?

Before we seek the possible alternatives, it will help us to take an overall look at the way man gets and uses energy today.

The six distinct 'forms' of energy we have discussed are:

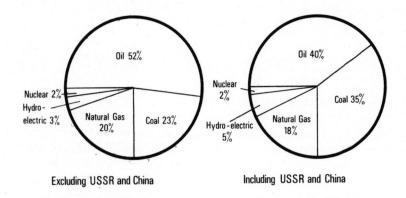

Excluding USSR and China   Including USSR and China

Fig 5. World energy consumption by sources (based on 1970 estimates)

1. Mechanical energy,
2. Heat energy,
3. Chemical energy,
4. Electrical energy,
5. Radiant energy,
6. Nuclear energy.

We have seen already that these are not altogether comparable. Mechanical energy, heat, electrical and radiant energy are examples of energy in motion, while chemical and nuclear energy comprise stored energy. Although this classification is useful it gives no hint of the origin of our energy. The diagrams shown in Figures 1 and 2 partially answered this question of origin, since they traced, in the first case the source of the fossil fuels, and in the second of water and wind power, to the radiant energy of the sun. To complete the picture we must locate the origins of geothermal energy, of tidal energy, and of nuclear energy.

The diagram on page 47 (Fig. 6) attempts to cover the entire field. The most significant feature of the diagram is the fact that *all* man's energy comes, originally, from the conversion of basic matter.

There are two fundamental processes at work — nuclear fusion and nuclear fission. Fusion takes place in the sun, from which all our chemical energy, as well as hydro-electric,

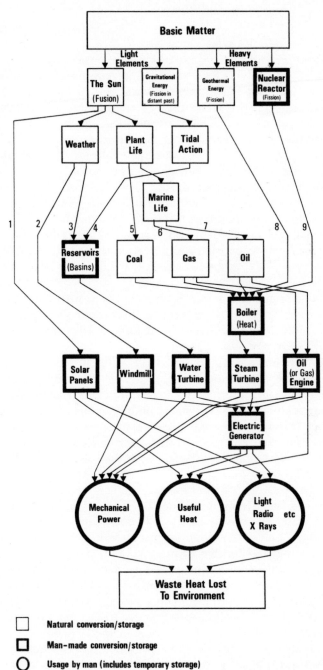

Fig 6. The energy circuit: basic matter to waste heat

Natural conversion/storage

Man-made conversion/storage

Usage by man (includes temporary storage)

wind-power and direct solar energy, are obtained.

Tidal energy is produced by the combined gravitational pull of the moon orbiting the earth and of the sun as the earth spins; but this is not original energy. While scientists are not all agreed on the origins of the solar system, there is little doubt that it was formed, at the beginning of time, as we know it, by some kind of nuclear explosion. It is therefore fairly likely that the energy of the moon's orbit came, originally, from nuclear fission.

Radioactive decay of unstable heavy elements is generally agreed to be the source of the earth's internal heat — the heat of which we see evidence in volcanoes, and which we have begun to use in geothermal power stations. This heat is the result, therefore, of nuclear fission which takes place slowly but steadily under the earth's crust.

The man-made nuclear reactor is the fourth basic energy converter, and here it is once again nuclear fission which is responsible for the conversion of basic matter into energy — mostly heat.

Though not shown in the diagram, scientists are at work trying to find a means of reproducing the hydrogen fusion process as an alternative to the nuclear fission reactor. Unfortunately there are many immense problems in maintaining this reaction under control (the hydrogen bomb, of course, involves an uncontrolled reaction) and there seems to be little real likelihood of success, at least not until well after the year 2000.

In any search for new energy 'sources' we first eliminate the fossil fuels (eliminate lines 5, 6 and 7 in the diagram). Wind, water (including tidal) and geothermal energy (lines 2, 3, 4 and 8) all offer scope for increased exploitation, though this is somewhat limited in practice and, since they are used almost exclusively for the production of electricity, do not offer a solution to the problem of replacing oil — at least not until a form of compact energy storage far more efficient than the electric battery as we know it, has been developed.

This leaves only solar and nuclear energy.

In theory it might be possible for either of these to be developed for use in transport. But in both cases there are serious problems. To collect sufficient solar heat (even assuming there are no clouds) solar collectors have to be too big to operate electric motors of the size and power output needed to drive a small automobile, let alone large trucks, railway locomotives, or aircraft. An average family car, for example, would need an absolute minimum of 25 square metres of solar panel area in constant bright sunshine to provide adequate power.

So far as ships are concerned the answer could very well be nuclear power. The American Navy's first nuclear powered submarine, the *Nautilus,* launched in 1954, proved the practicability of nuclear power for ships on its maiden voyage, when it travelled 100,000 kilometres without refuelling.

A conventional diesel powered submarine would have consumed about ten million litres of oil to cover the same distance. The *Nautilus* and the nuclear powered merchant ship the *Savannah* (put into service five years later) both proved another point — that the great weight of shielding necessary to protect the ship's crew and cargo from radiation, in case of accident, was not a serious handicap. In the *Savannah,* a vessel of 22,350 metric tons displacement, the 'biological' shield along with a seven ton load of uranium fuel weighed a little over 2,000 metric tons. An oil-fired merchant ship of this size would normally carry up to 3,000 metric tons of oil (or ballast in lieu) and this would power the 22,000 horse power engines for a small fraction of the half a million kilometres obtainable from the seven tons of uranium.

The Russians made use of this long range potential of the nuclear powered ship for the *Lenin,* an ice-breaker with high powered engines. It was designed to lead merchant convoys through the Arctic sea route north of Siberia. By using uranium as its fuel the *Lenin* was able to stay at sea for periods up to a year, whereas an oil-fired ice-breaker would have lost considerable time in returning to port for refuelling.

The simple fact is that one gram of uranium-235 'consumed' by the fission process, produces the same heat output as 3½ metric tons of coal. Weight for weight it releases 3½ million times as much energy.

For reasons which will be explained in Chapter 9 the

nuclear reactor has to be relatively large if the nuclear chain
reaction is to be both sustained and kept under control. Add
to this the weight of the radiation shielding needed, and the
reactor is virtually ruled out as a power source for land or air
transport. Heat from natural radio-active decay not involving
a chain reaction can be made to produce an electric current,
but in this case the energy output can never be sufficient to
power even the smallest car. The process has been utilized in
cases where a long-term and reliable low-output energy
source is required, such as in the so-called heart 'pacemaker',
which will operate inside the human body for about ten years
without refuelling. It has also been used in some navigational
buoys.

Having more or less eliminated the currently proven means
of energy supply for small-scale transport it seems clear we
must either find an entirely new 'mobile' fuel to replace oil,
or develop means of storing energy in a far more concentrated form than is currently possible with the lead-acid
battery.

The idea of finding a new fuel is a non-starter. To be a fuel
a substance must be something in which energy has been
stored by a natural process. The fossil fuels qualify because
they are found in substantially the same form as that in
which they will be used. The energy stored in them has been
supplied by the sun. Hydrogen is sometimes proposed as a
potential mobile energy source. Hydrogen is certainly a very
common element because it is one of the constituents of
water which exists in vast quantities. But the quantity of free
hydrogen available in the world is small, and to obtain
hydrogen from water in the form we need we have to use
energy to secure the separation of the hydrogen from the
oxygen. In other words the element hydrogen can only act as
a means of *storing* energy which we first put in and then take
out again. We shall take a look at this possibility in Chapter
10.

We have already spoken of another possible field for
energy storage research — that of developing means of storing
electricity in a more concentrated form than is today possible
with the lead-acid battery. This, too, will be discussed in
Chapter 10.

If the world is today as dependent on the fossil fuels as I have made out, the statistics of their consumption by the various countries, of world resources, and of the known potential of other energy sources provides a great deal of food for thought.

To present these statistics simply and clearly is not easy because coal, oil, gas, hydro-electric energy and nuclear fuels are all measured in units which are not directly comparable (for example we have tons of coal, barrels of oil, cubic metres of gas, watt/hours of hydro-electric energy, and tons of uranium *oxide* − not of uranium *metal*). Moreover published statistics are not always complete or up-to-date on a world wide basis. For this reason the analyses I give below are based on approximate percentages in terms of the heat outputs of the various fuels. While available statistics refer to various years I have attempted to give all percentages corrected for 1970 by adjusting available figures in keeping with known trends. These should not have altered significantly until the changes which followed the 1973 Arab-Israeli war, and the subsequent controls on oil output and the price rises imposed by the Middle East producers. Some of the figures are, however, changing as a result of recent discoveries in oil reserves.

Diagram A (Fig. 7) shows the distribution of world consumption of coal and Diagram B shows the distribution of

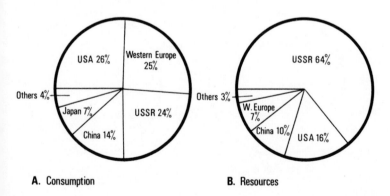

A. Consumption   **B. Resources**

Fig 7. World consumption and resources of coal (based on 1970 estimates)

the world's estimated reserves (which are thought to total about nine million million metric tons). The figures show that the U.S.S.R. and China are well off. While they consume about 38 per cent of the world's coal they possess about 74 per cent of the world's reserves. The largest single user of coal, at 26 per cent, is the United States, but it holds only 16 per cent of the world's reserves. Western Europe is still worse off. It holds 7 per cent of the world's reserves, but currently accounts for about 25 per cent of world consumption.

Before we can make comparisons in the field of oil we must decide whether we are to include in oil reserves the potential of the tar sands of Canada and of the tar shales of the United States, Brazil and elsewhere. As the tar sands and shales had not been commercially exploited up to the time of the 1973 oil crisis, it seems logical first to exclude these sources of oil from our view. Diagram C shows how oil

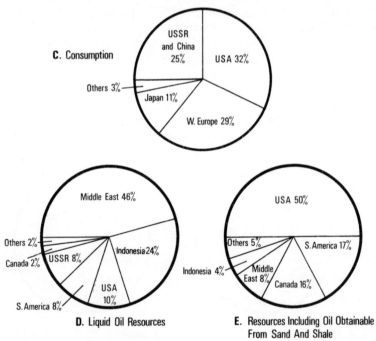

Fig 8. World consumption and resources of oil (based on 1970 estimates)

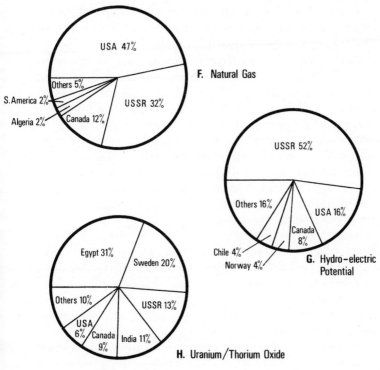

Fig 9. *World resources of natural gas, hydro-electric potential, nuclear fuel (based on 1970 estimates)*

consumption is divided around the world and Diagram D shows where this oil has been coming from. We can see at once that while the United States has been consuming about a third of the world's oil production it only holds about 10 per cent of the world's reserves. Western Europe which has been consuming only a little less than the United States, and Japan, which has been accounting for about 11 per cent of world consumption, together have virtually no reserves of oil. On the other hand the Middle East countries were estimated, about 1970, to possess 46 per cent of world reserves, Indonesia having 24 per cent, United States and South America between them 18 per cent and Russia about 8 per cent. A comparison of Diagrams C and D shows immediately how it was possible for the oil crisis to arise, especially when

one remembers that all the oil represented here is only
expected to last the world for about 30 years. (Recent
discoveries have increased this figure.)

Now take a look at Diagram E which shows the reserves of
oil including that obtainable from the Canadian tar sands and
the tar shale in the United States, South America and
elsewhere. Now we see America as possessing 50 per cent of
the world's reserves of oil, very considerably more than its
percentage consumption. Canada holds a further 16 per cent
whereas it currently accounts for only 4 per cent of world
consumption. In this diagram the Middle East holds only 8
per cent of world reserves and Indonesia only 4 per cent.
South America, because Brazil has fairly extensive deposits of
tar shales, has its percentage pushed up to 17.

The implications are clear. If the United States had
developed the extraction of oil from tar sands and shale there
would be no oil crisis today. The reason they did not do so
was because oil from these sources would not have been able
to compete economically with Middle East oil. No wonder
the Arabs decided, with their own oil rapidly running out, to
put up their prices.

Diagram F shows the distribution of the world's reserves of
natural gas (totalling over seventy million million cubic
metres). As natural gas usage broadly follows the same
distribution pattern, I have given no corresponding diagram
for demand. Part of Russia's share is exported to Europe,
Italy being an important customer.

The same is true of hydro-electric energy exploitation.
Obviously hydro-electric power stations can only be built
where the potential actually exists, so Diagram G, which gives
the distribution of potential hydro-electric sites in terms of
potential power output, tells us all we need know. We may
add, however, that the United States has already exploited a
far greater part of her potential than has Russia.

The use of nuclear power is at present so small a percentage of the world's total energy consumption that here,
once again, I have only given the distribution of known
deposits of uranium and thorium oxides, the raw materials
from which nuclear fuels are obtained. But here we see at
once some politically thought-provoking facts. The United

States, which is the world's greatest consumer of energy, holds only 6 per cent of the world's known reserves of nuclear fuel. Western Europe holds even less. Canada and India seem fortunate in possessing 9 per cent and 11 per cent respectively of the world's uranium and thorium deposits. Soviet Russia has 13 per cent, a useful reserve, but note the strong positions of Sweden with 20 per cent and Egypt with 31 per cent of the world's uranium and thorium reserves. With oil reserves running dangerously low, and nuclear energy the likely replacement for fossil fuel energy in the immediate future, Diagram H is worth thinking over. Yet even the national distribution of uranium and thorium deposits could be unimportant if the price of energy continues to rise. Sea water contains salts of natural uranium which can be extracted at a cost about 25 per cent higher than the current price of uranium oxide. And there is enough uranium in the oceans to supply all man's foreseeable energy needs for hundreds of thousands of years.

CHAPTER 5

The Conversion of Chemical Energy

The animals of pre-history learnt to store and make use of chemical energy long before the intelligence of certain species became developed to the point where the first man was 'created'. For every time a living organism or being absorbs 'food', it is putting 'fuel' into its system.

The slimming woman is all too conscious of the energy content of various foods. A slice of bread, she is told, contains so many calories. We have seen that a calorie, in physics, is the quantity of heat needed to raise the temperature of one gram of water by one degree centigrade. A food Calorie (note the capital letter) is what the physicist would call a kilocalorie. In other words 1,000 calories equals one food Calorie. So one food Calorie is the heat needed to raise the temperature of 1,000 grams of water by one degree centigrade. If a slice of bread yields 100 Calories, this means that, fully 'burnt', that slice could produce enough heat to raise one kilogram of water from freezing to boiling point.

The 'fuel' one eats is not fully burnt in the body. To understand this we must see how energy gets into food. Photosynthesis, the process which stores the sun's radiant energy in the form of chemical energy, splits the water atom, freeing the oxygen and combining its hydrogen with carbon dioxide from the air. The result is a carbohydrate, the basis of all vegetable matter. Carbohydrates are found in many forms of which the most common, in food, are starches and sugars.

One of the processes which takes place in the bodies of animals and men, is the conversion of the chemical energy of sugar into the mechanical energy of contracting muscle. While this process is far more efficient than that of the steam engine (30 per cent of the fuel's chemical converted into mechanical energy), of the petrol engine (50 per cent efficiency) or even of the electric motor (90 per cent efficiency), some of the chemical energy is converted into heat, which warms the body. This explains why physical action warms you.

Other body processes convert basic carbohydrates into fat or protein, the former by freeing some of the oxygen, the latter by a more complicated process which requires the addition of other elements. In both cases most of the chemical energy of the sugar or starch remains stored. While protein is used for body building, fat is made simply as a means of storing surplus energy. In fact fat, which contains a smaller percentage of oxygen than sugar, is a more concentrated store of chemical energy. To form it from sugar some of that sugar is 'burnt' (i.e. converted into carbon dioxide and water) to provide the additional energy. When the body is 'hungry', needing more energy than is available from the starch and sugar in its food, it makes use of some of its stored fat, breaking it down to release the energy it needs.

Before man discovered fire he had relied for warmth partly on the conversion of solar (radiant) energy into heat, and partly on the heat produced as a by-product of carbohydrate conversion in the body. When he first learnt to control fire he had opened up a completely new world of energy conversion. He had found a way of converting stored chemical energy directly into heat.

Today this process of burning is the basis of most of man's methods of converting stored chemical energy into other forms which are more convenient for his use. We warm ourselves, our houses, our offices and our factories mostly by burning wood, coal, gas or oil. We produce mechanical energy by reconverting the heat of fuel combustion. We produce electricity by once again reconverting mechanical energy.

Not only are carbohydrates and hydrocarbons (the latter have no oxygen in their molecules) used as fuels for the production of heat. Hydrogen, for example, has been used along with oxygen, both in liquid form, as a rocket fuel. The hydrogen is simply burnt, producing water. The great quantity of heat energy released expands the water vapour enormously, producing the thrust that drives the rocket. Good quality coal is almost pure carbon which burns with oxygen to form carbon dioxide. (If the oxygen supply is insufficient the poisonous gas carbon monoxide is first formed, less heat being released; but this gas is also easily burnt, when more oxygen becomes available, giving out more heat during its conversion to carbon dioxide.)

Methods for securing the release of heat energy from the fossil fuels are many. They range from the simplest to the very complex. The ordinary coal fire, the gas fire and the gas cooker, require no explanation. The central heating boiler, whether it burns coal or gas or oil, is almost as simple though the oil boiler needs a means to 'atomize' the fuel to enable it to burn more quickly.

Four-fold Energy Conversion

Since electrical energy today accounts for at least half of the world's total energy consumption and stored chemical energy (in the form of the fossil fuels) provides most of our current supply, it follows that the most important man-made energy conversion process is the conversion of chemical into electrical energy. Some forms of stored chemical energy can be converted directly into electricity — the electric battery and the fuel cell do so — but this is not true of the fossil fuels. Here the conversion process used widely around the world in the conventional thermal power station is a four-stage process. First the fuel is, literally, burnt so as to release heat. So we have chemical energy converted into heat energy. Next we use the released heat to raise the temperature of water until it boils producing a pressure in the boiler, and then to further raise the temperature of the steam in order to increase that pressure. We now have our heat energy converted into stored mechanical energy. The steam is now led through a turbine — a device which converts the potential energy of the steam into rotary kinetic energy. Finally this mechanical energy is used to drive a generator, which converts it into electrical energy.

Fuel into Heat

The release of heat from coal, oil or natural gas presents no problem. We all know how to light a coal fire, an oil heater and a gas stove. Once the material begins to burn, the chain reaction maintains the process, providing the fuel and sufficient air (containing the necessary oxygen) are supplied. One kilogram of coal, fully burnt, will surrender over 7 million calories of heat. Paraffin's heat output is better, rising to almost 10 million calories per kilogram. Natural gas performs best of all; only here we have to face the problem that a kilogram of gas occupies a great deal more space. Ten kilograms of any of these fuels is sufficient to heat an average room for an hour or more. The coal's heat, for example, is equivalent to the output of a 1 kilowatt electric heater burning for 60 minutes.

Compare with this the output of a large modern thermal power station, which may be rated as high as 500 million watts (500 MW) or more. To achieve this output of electricity the daily consumption of coal will be about 5,000 metric tons, or over 200 metric tons an hour at peak periods. Not only does 200 tons of coal have to be fed into and burnt in the furnaces every hour, but about 750 metric tons of ash have to be removed from the furnaces, and from the power station area each day. And, of course, the 5,000 tons of coal has to be brought into the station's coal storage area each day. The main problem is one of what we call logistics — the science of transportation.

Coal will arrive at a power station loaded on railway wagons, and the first step in easing the problem is to locate the power station as close as is possible to both a coal mining area and the area where the electricity is needed. The transmission of electricity is not a problem of moving heavy loads, like coal; but electric cables have electrical resistance, and the shorter the transmission lines the less the electricity lost as wasted heat.

In a typical large power station the incoming wagons of coal are first marshalled into sidings so that, even if some trains are held up by weather or locomotive failure, or even by strikes, there will always be wagons waiting for unloading. Wagons are unloaded automatically by machines which tip the coal on to an electrically driven conveyor belt. This carries it to a storage area where as much as 100,000 metric tons of coal may be stockpiled in a large power station.

Further conveyors carry the coal from the stockpile to huge coal crushers, which reduce the fuel to small lumps. From these crushers the coal moves on, again by conveyor belt, to the furnace houses, where it is now passed through pulverising mills which reduce it to a fine powder. Finally

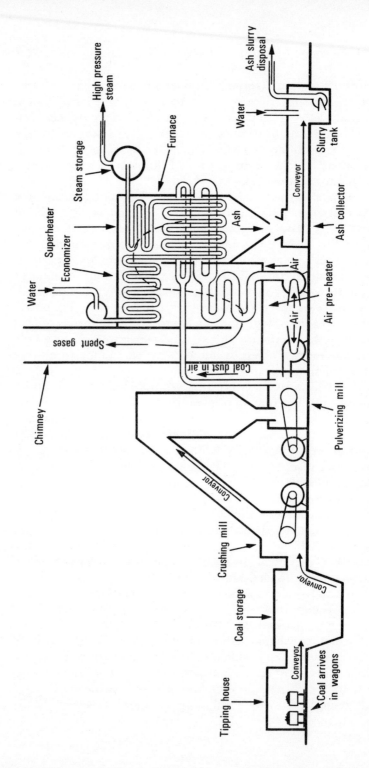

Fig 10. Coal supply and furnace system at typical power house

this coal dust is caught up in a current of air which carries it through large pipes straight into the furnace. There it burns fiercely and rapidly with long swirling flames, additional pre-heated air being supplied by huge centrifugal fans.

Heat into Pressure

Inside the furnace is an array of steel pipes through which water is constantly pumped. The intense heat boils this water rapidly, producing steam under pressure, and the water flow is regulated so that the steam temperature, pressure and quantity meet the requirement of the turbines into which it is fed. Typical figures are 500 metric tons of steam each hour at a temperature of about $550°C$ and a pressure of 1,700 kg/m^2. To obtain this high level of temperature and pressure none of the available heat must be wasted. To ensure full usage of the heat two features are included in the steam pipe layout. In the 'superheater' the pipes pass finally through the hottest part of the furnace flames, near their tips, before leading out to the turbines; and in the 'economizer' the spent gases, from which much of the heat has already been lost to the steam, are used to warm the incoming water and air before being released into the chimney. (*See* Fig. 10).

The ash from a coal furnace falls on to a conveyor at the bottom. This carries it away to huge bunkers where it is mixed with water to form a slurry which can be pumped away to a disposal area.

Some power station furnaces burn oil or natural gas. In the latter the gas is fed in under pressure, a separate air supply, also under pressure, being provided. In oil furnaces the oil is introduced in the form of a fine spray carried by air under pressure. In both cases, provided the correct quantity of air is pumped in with the fuel, complete combustion ensures the maximum heat output. In some modern power stations the furnaces are designed to burn either coal or natural gas, or either coal or oil. This provides flexibility which enables the furnace to remain in action in case of a fault in fuel supply plant or during temporary fuel shortages.

Heat into Mechanical Energy

The modern steam turbine was invented in 1884 by Charles Parsons, a 30-year-old engineer and junior partner of an engineering company at Gateshead, England. Parsons needed an efficient engine to drive dynamos at his company's factory and realized that the inefficiency of the early steam engine was largely due to the relatively slow oscillating system of pistons sliding in cylinders, and to the friction of the many associated working parts. Other men had experimented with steam turbines built like multi-bladed electric fans. These proved extremely inefficient in practice and Charles Parsons realized that the steam passed out from the back of the blades with almost as much energy as it brought to it. Only the relatively small difference in incoming and outgoing energy was being converted into rotary motion by these blades. Parsons' idea for overcoming this was a simple one. He built a series of multi-bladed discs and fixed one behind the other on a single shaft. He reasoned that the steam would give up some of its energy as it pressed against the blades of the first disc; passing on at a slightly reduced pressure it would come to the second disc where the action would be repeated, more energy being transferred to the shaft. A third disc would again repeat the process. To secure the maximum effect of the steam pressure on the blades of each disc, Parsons fitted a disc of fixed blades in front of each rotating disc. The fixed blades were set at an angle so as to deflect the steam to hit the moving blades roughly at right angles. The steam deflected off these rapidly spinning blades would then meet another set of fixed blades, and the action would be repeated.

Parsons' first steam turbine, based on this principle, used steam at a pressure of 575 kg/m^2. The rotor, spinning at 18,000 revolutions a minute developed about 10 horse power.

Parsons, who was later knighted, subsequently made improvements to his turbine design, the most important being to vary the size of the blade discs to match the expansion of the steam, and the corresponding lowering of its pressure as it passed through the turbine. At the inlet end, where the steam entered under quite high pressure, the

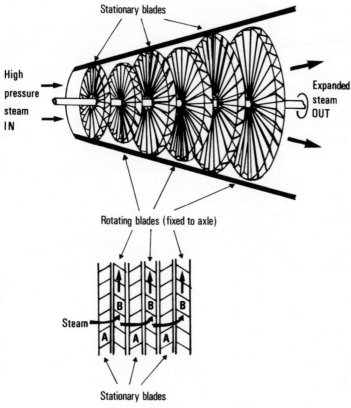

High pressure steam IN

Expanded steam OUT

Stationary blades

Rotating blades (fixed to axle)

Steam

Stationary blades

Fig.11 Basic design of steam turbine

diameter of the first set of moving blades was small. At the other end, where the spent steam emerged at low pressure (much of its energy having by now been transferred to the spinning shaft) the diameter of the final set of blades was much larger. This design improvement greatly increased the efficiency of the turbine and today most modern steam turbines have three separate sets of blades of varying diameter, for high pressure, medium pressure and low pressure steam. The turbine discs in each of the three units also increase in size from inlet to outlet. (Fig. 11)

Mechanical Energy to Electricity

The link between electricity and magnetism was first observed in 1819 by a Danish physicist, Hans Oersted. He was demonstrating a galvanic battery to students at Kiel University when he accidentally let fall a wire, through which an electric current was flowing. The wire fell across a mariner's compass lying on the table and when Oersted looked down he noticed that the compass needle was no longer pointing north. When he lifted the wire off the compass its needle swung back to its normal north-south position. The explanation was that an electric current flowing in a wire creates a magnetic field around it. This field was reacting with that of the magnetic needle. When an electric current is passed through a coil of wire the entire coil produces a magnetic field similar to that of a bar magnet, lying along the coil's axis, with a north pole at one end and a south pole at the other.

Oersted's discovery led to the invention of the electric motor. In its simplest form this consists of a wheel, around the circumference of which are fixed a number of compact coils of wire. The wheel is mounted between the poles of a magnet. An electric current is passed through each of these coils in succession so that a coil approaching either of the magnet's two poles is energized to produce an 'opposite' pole on the same side. As unlike poles attract one another the wheel is turned until the opposite poles are as close as is possible. At this point the electric current is transferred automatically to the next approaching coil. The current again produces an 'opposite' pole which is drawn towards the pole of the magnet. By making the turning wheel itself switch the current from coil to coil the process is synchronized and the wheel spins faster and faster.

Just as an electric current flowing in a coil produces a magnetic field, the passage of a coil of wire through a magnetic field causes an electric current to flow in the wire. The simple electric motor I have described can therefore be used 'the other way round.' By turning the wheel the coils on its circumference pass through the magnet's field turn by turn. This causes an electric current to flow in the wire of each coil in succession and the automatic switching system leads this current out of each coil, turn by turn, resulting in the production of a continuous current. This is the principle of the electric generator.

If, instead of a series of coils and a switching mechanism, we have a single coil on the wheel and measure the electric voltage produced in the wire as the wheel turns, we will find that this rises, as the coil approaches one of the magnetic poles, and then falls to zero as it passes on. As the wheel continues to turn the coil will now approach the magnet's other pole. This time the voltage will again rise, but due to the opposite magnetic polarity the voltage will also be of opposite polarity. So if the wheel continues to turn the voltage across the coil will oscillate from zero to a maximum point, and then back through zero to a minimum (or *negative* maximum) point; it will then return to zero once more. This is what we call an alternating voltage. When wires carrying an alternating voltage are connected to a circuit the current that flows is an alternating current (AC), flowing first in one direction and then in the other.

If, instead of one coil, the generator wheel has three separately connected coils spaced equally around its circumference, the result is a series of three overlapping alternating voltages. The voltage peaks follow each other in succession. If these three outputs are fed into the three wires of a three-core cable, we have what we call a 3-phase AC supply.

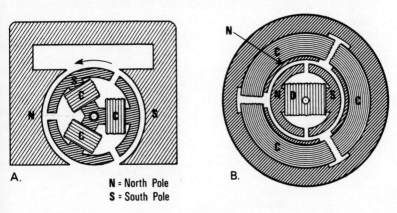

A.

N = North Pole
S = South Pole

B.

Fig 12. Principle of electric generator

The principle behind the operation of the electric generator is based on the movement of coils of wire through a magnetic field (Fig. 12). In practice it is not essential that the coils should move and that the field should be stationary. If the coils are still and the magnetic field moves the result is exactly the same. This alternative makes the construction of large AC generators considerably simpler. The three main coils, in which the alternating voltage will appear, are fixed coils wound so that they lie flush on the inside of a huge cylinder. Because they do not move these three sets of coils, one for each output phase, are together called the 'stator'. The cylindrical space inside them, which may be a metre or more in diameter, accommodates the 'rotor' of the machine. In theory the rotor could be a huge cylindrical permanent magnet. In practice it is an electro-magnet with its own coil through which direct current is passed to produce the required magnetic field. The direct current needed is generated on the spot by means of a relatively small dynamo fitted on to the end of the rotor. It is known as the exciter.

The complete AC generator (called an alternator for short) is coupled direct to a high-speed turbine so that it rotates at the turbine's speed. It is a huge piece of machinery. A typical alternator weighs 1,000 or more metric tons and produces a 3-phase electrical output at voltages up to 16,500. The power output may be as high as 100 MW or more. This is a great deal of energy — enough to supply 100,000 single bar (1 kW) electric fires all switched on at once.

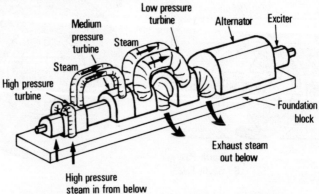

Fig 13. General layout of complete turbo-alternator

CHAPTER 6

The Wind and the Rain

The wind and rivers have been used as sources of energy for centuries. Windmills were used all over Europe to pump water and to grind wheat, and around the world the wind drove the sailing ships on which world trade once depended. Why else were trade winds so called?

The water wheel, too, has long been used for the conversion of river flow into mechanical power. By the nineteenth century engineers had designed water wheels which were remarkable by any standard. A pair of 16 metre diameter water wheels, each 3.5 metres wide, installed at a Scottish cotton mill in 1824, developed 500 horse power from a modest 16 metre head of water. The efficiency of energy conversion was 75 per cent; and though these wheels only turned at three revolutions a minute they were geared to operate machinery which included parts rotating at 9,000 rev/min.

Wind Power

We saw briefly in Chapter 1 how winds are created partly by the inertia of the air above the constantly spinning earth, and partly as a result of the radiant energy of the sun warming the equator more than the poles and the land masses more than the sea. The energy of all the winds that blow steadily over land can never provide more than a fraction of the modern world's needs. But with our fossil fuel reserves diminishing fast should any significant source of energy — especially free energy — be left untapped?

There have been ambitious plans to harness the wind. Between the wars there was one such plan in England. (Doubtless there have been others elsewhere.) The English plan was to build a network of wind-powered electric generating stations to make use of the Westerlies, the winds that blow reasonably steadily in those parts of the British Isles where high ground does not disturb the natural weather pattern. An experimental wind power station was actually

built in one of the Orkney Islands. But the scheme was
abandoned in the 'fifties following the advent of nuclear
power which, it was then thought, would quickly meet all
Britain's energy needs at a competitive price.

When the electric motor was invented, in the mid-
nineteenth century, and the conversion of electrical into
mechanical energy became a practical proposition, the
storage battery was, at first, the energy source.

The electric motor can, of course, be used the other way
round, to convert mechanical into electrical energy, and
when the first generators were put to practical use it was the
steam engine which was used to drive them and coal was the
fuel. Then a German engineer had a brainwave. Why should a
generator not be powered by a water wheel? The result, in
theory, would be free electricity. His idea was demonstrated,
though little noticed, at the 1882 Munich Exhibition, where
an electric motor was kept constantly running.

Water Power

The weakness of the exhibit was that the source of the
motor's energy could not be shown. In fact the motor was
connected by copper wires to a similar motor installed on the
bank of the river Isar at the town of Hirscham, about 10
kilometres away. The second 'motor' was driven by a water-
wheel. It worked as a generator.

In the hydrological cycle the sun's energy is converted
naturally into the potential energy of vast quantities of water
suspended in minute droplets forming clouds high in the sky.
These clouds produce, as a result of rainfall, a constant
energy flow in the world's rivers of about 80 million million
watts. This is substantially more than half of mankind's total
energy demand today. Even a single major river dissipates an
enormous quantity of wasted energy. For example the energy
lost as the water of the Zambezi river flows over the Victoria
Falls, where 100,000 cubic metres fall just over 100 metres
every second, is approximately 1,500 megawatts.

In modern practice the simple water wheel has been
replaced by the turbine which is powered, in most cases, by
water fed through pipes. There are two basic designs known
as the impulse turbine and the reaction turbine.

When water held by a dam is allowed to flow, the water's potential energy is converted, by the force of gravity, into kinetic energy. The potential energy of the water in the reservoir is a function of the mass of the water and its height above sea level. (We use mean sea level as the basis of measurement as all water raised by the action of the sun's energy will ultimately, if not diverted, end up in the sea.) We cannot use all the potential energy of a reservoir, as we cannot normally site a power station at sea level. We are therefore concerned with what we call the 'head' of water — that is the difference in level between the surface of the reservoir and the turbines in the power station.

In practice water is usually 'piped' from a reservoir to a turbine (a tunnel through mountain rock is equivalent to a pipe). The water inside the pipe is under pressure, and this pressure depends on the 'head' of water above it. But when it reaches the turbine it can be used in two distinct ways. Imagine that the water supply pipe is led to a nozzle like that on a fireman's hose. The water will emerge in a powerful jet in which the water velocity is higher than in the pipe. As the water is no longer enclosed in the pipe it will be at atmospheric pressure. If this jet is directed on to the blades of a fan, these blades will turn. The pressure of the water on the blades of the fan depends on the size of the jet and the speed of the water in it. This is the principle of the impulse turbine.

Now imagine a ship's propeller fitted snugly inside the pipe supplying water from a reservoir. If the fit is a good one the propeller will act as an obstacle to the passage of water, and the water pressure in the pipe, acting on one side of the propeller blades, will produce a 'reaction' which turns the propeller. This is the principle of the reaction turbine.

Each of these two types of turbine has advantages over the other. A modern impulse turbine will convert as much as 90 per cent of the water's kinetic energy into rotary motion, provided high velocity water jets are used. To produce a suitable jet the head of water must be high and in practice the impulse turbine is rarely used where the head is less than 200 metres. It is found at its best when the head is 1,000 metres or more. When the water flow in the jet is reduced the power output of the impulse turbine is reduced roughly in proportion. It is therefore flexible in use, and can be made to respond quickly to fluctuations in the 'load' it is being used to drive. (We shall explain this idea of 'load' later.)

The best reaction turbines are no less efficient when operating at the water pressure for which they were designed but their power output falls rapidly if the pressure changes. They are therefore less adaptable to situations where the load is likely to vary, though they can, of course, be easily started or stopped by operating water valves. On the other hand the reaction turbine is efficient at very much lower heads than the impulse turbine; there are, in fact, two distinct types of modern reaction turbine. One is most efficient with heads between 30 and 1,000 metres at the most; the other is used mainly in situations where the head of water is less than 50 metres. The latter type is efficient at heads of only 7 or 8 metres, and will even continue to work at heads as low as 1 or 2 metres, though its energy conversion efficiency drops rapidly at heads under about 5 metres.

The Impulse Turbine

In early impulse turbines the water jet was directed on to a series of flat vanes. Only 40 per cent of the water's energy was converted into rotary motion, a large percentage being absorbed by splash. The introduction of cupped vanes raised the efficiency to about 65 per cent. Then in 1889 Lester Pelton, an American engineer, designed an impulse turbine which had vanes in the form of twin cups. The jet played on the sharp junction of the cups. This divided the water between the two cups and the efficiency rose to 80 per cent. The twin cupped impulse turbine became known as the Pelton Wheel and later alterations in the cup profile, and improvements in the nozzle which formed the jet, achieved an efficiency of 90 per cent. (*See* Fig. 14).

The modern Pelton Wheel, which is operated with two, four, or even six jets playing on it simultaneously, is widely used in high-head power stations.

The Reaction Turbine

One of the earliest reaction turbines was an inefficient device

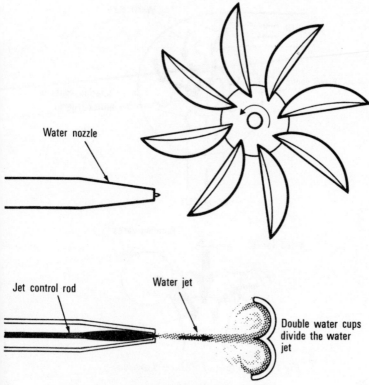

Water nozzle

Jet control rod

Water jet

Double water cups divide the water jet

Fig 14. Principle of Pelton Wheel

known as Barker's Mill. Invented in England in the 18th century it worked on the principle of a modern spinning lawn sprinkler. Water was fed, under pressure, into the axle of the spinner, from which it emerged through a pair of nozzles fitted tangentially on opposite sides and pointing in opposite directions. The reaction of the jets acted to turn the spinner; but as much of the water's kinetic energy remained in the emerging jets, only a small proportion was converted into rotary motion.

In 1827 a Frenchman called Fourneyron designed a turbine in which the water was fed under pressure into the centre of a squat cylindrical container having slots around its circumference. The water emerged radially from these slots to press on the vanes of an annular runner rather like the air

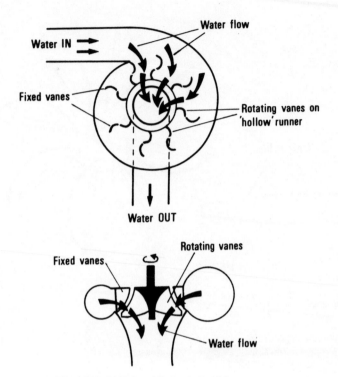

Fig 15. Principle of Francis Turbine

wheel of a modern electric turbo-heater; 75 per cent efficiency of energy conversion was achieved. In 1855 an American, J. B. Francis, did better by building a similar turbine 'inside out'. In the modern Francis Turbine the water, under pressure, is fed into the large end of a spiral 'volute' chamber (which looks like a huge snail's shell). The water passes inwards through a series of fixed vanes which guide it on to the curved vanes of a runner rotating within.

The efficiency of the modern reaction turbine depends greatly on the angle of the runner blades, and it was an Englishman, James Thompson, who improved the Francis Turbine by fitting adjustable runner vanes.

In 1913 a Czech engineer patented an entirely new type of reaction turbine. which carries his name, Kaplan. The modern Kaplan Turbine, which operates efficiently at heads as low as

10 or even 5 metres, looks like a miniature submarine with one huge propeller at one end. The turbine is fitted in a large diameter water pipe, which may be vertical or horizontal, and efficiency is improved by fixed guide vanes fitted ahead of the rotating runner vanes. These guide vanes deflect the flowing water so that it meets the runner vanes at the angle which produces the maximum energy conversion. For control of the energy output the runner vanes are designed so that they can be 'feathered' from within the hub (Fig. 16).

As most hydro-electric turbines rotate on a vertical axis, the alternator, which is fitted above, is of rather different design to that used in thermal power stations. We saw in Chapter 5 how the alternator is normally designed to give a three-phase output, with the peaks and troughs of the alternating voltage staggered by a time lapse equal to one third of a cycle. (The AC cycle in Great Britain in 50 Hz — one Hertz is one cycle per second — while that in the U.S.A. is 60 Hz.) In our explanation of the working of a three-phase alternator (page 65) we showed three large coils spaced around the circumference of the stator. To produce a 50 Hz 3-phase AC supply this means that the rotor must turn at exactly 50 revolutions each second, or 3,000 rev/min. As a reaction turbine revolves much slower the 50 Hz output is achieved by having a correspondingly larger number of coils on the stator. If, for example, the runner of a low-head reaction turbine is designed to turn at exactly six revolutions each second the alternator must have 25 coils on its stator, to give a 50 Hz 3-phase output, (6 x 25 = 150 voltage peaks a second; of which 50 go to each phase each second.)

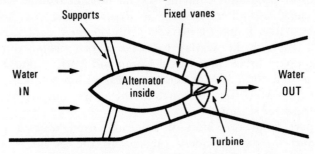

Fig 16. Kaplan Turbine with bulb alternator

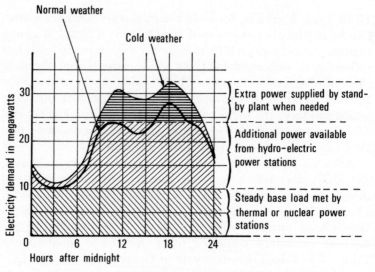

Fig 17. Typical electricity demand pattern for winter day

The Flexibility of Hydro-Electric Energy Output

The energy output of a large thermal power station cannot be quickly altered. It takes time for steam pressure to be built up in a furnace and it is uneconomical to maintain surplus high pressure steam to enable an idling turbo-alternator to be brought rapidly up to full power in response to a sudden demand. In general terms electricity demand over the 24 hours of the day follows a steady pattern of rise and fall. But a simple event, like a commercial break in a highly popular televison programme having several million regular viewers, can cause a sudden upsurge in demand. Suppose, for example, half a million housewives put on their electric kettles (each using 3 kilowatts) to make a pot of tea during a commercial break in a Royal Command Variety Show. This means that the national demand rises, within a few seconds, by 1,500 MW. This is equal to the entire output of three major power stations.

Apart from this kind of situation, even 'normal' power demand fluctuates widely throughout the 24 hours of the day. Minimum demand is between 1 and 6 a.m., while maximum is from 8 a.m. to 9 p.m. with high peaks at about 9 a.m. and at 5 p.m. As thermal power stations need to operate continuously at their designed output if the cost of the electricity produced is to remain competitive, this means that thermal power stations are not economical if they are the only source of electrical energy (*see* Fig. 17).

Hydro-electric power is, by contrast, extremely flexible. Power output is easily and economically cut by reducing the volume of water allowed to flow. Moreover a hydro-electric power alternator, which may be kept spinning with very little water flow, can be brought up to maximum output, in response to electricity demand, in a matter of a few seconds, merely by opening the sluices or valves which control the flow of water.

For this reason thermal power stations are generally used to maintain what is termed the 'base load' of the area they serve. To augment output during the high load period of a normal day hydro-electric power is ideal, with quick starting diesel generators kept in readiness to meet temporary peaks.

The 'load' on a power station is really another term for the demand. It is a measure of the energy taken from the alternators and is felt as magnetic resistance to their rotation. It is important that the load at any given moment is accurately balanced by the energy input, because if the input is less than the load, an electric circuit automatically 'balances its books' — in this case by a fall in alternator voltage. This, as we all know, is highly undesirable. Not only do electric lamps produce less light when voltage falls, but electronic equipment like television receivers, and appliances like refrigerators, may cease to work properly and can even suffer damage.

Pumped Storage

The conversion of the potential energy of water into electrical energy can easily be 'reversed'. For example if the base load supplied by a network of thermal power stations is greater than night-time demand in the area, surplus electrical energy can be used to operate powerful electric pumps which raise water from a low-level to a high-level reservoir. Then, when demand rises in daytime, the water pumped up during the night can be fed down again to power hydro-electric turbo-alternators. This system is known as pumped storage, and can be used to balance the books of electrical energy output and demand where the former remains constant and the latter follows a 24-hour pattern of rise and fall. A scheme of this kind in the state of Missouri, U.S.A., raises water 280 metres above the generators for seven of the 24-hour output of a 350 MW power station. This scheme uses a man-made high reservoir, which involves heavy capital cost. In North Wales a similar scheme makes use of a natural mountain reservoir at Ffestiniog, about 8 kilometres from and 325 metres above a 500 MW nuclear power station at Traws-fynydd. Four 75 MW electric pumps operate for about six hours each night using 300 MW of the power station's normal output. During peak demand periods next day four 80 MW hydro-electric generators are used for about 4 hours to produce a maximum output of 320 MW. Calculation shows that the pumping consumes, on average, a total of 1,800 MW hours of energy, while the re-generation produces only 1,280 MW hours. The 'loss' is due partly to the efficiency of both pumps and generators being less than 100 per cent, and partly to the resistance in the water pipes, met during both pumping and power generation.

CHAPTER 7

Energy from the Tides

The constant ebb and flow of the tides has engaged the attention of inventors for many hundreds of years. There is even a Domesday Book entry which describes a tidal mill at Dover. Many others are known to have existed along the English Channel, in both France and England, especially in the eighteenth and nineteenth centuries. Patents registered in both countries during the eighteen hundreds show a thorough understanding of the principles involved and describe a remarkable variety of tidal 'machines'. The simplest was in the form of a rotating paddle wheel on a floating mill, operating like a river water wheel; it required a site naturally channelled so as to cause a swift tidal flow. There was also the heavy raft which was lifted by the flowing tide and which fell on the ebb tide, to do useful work through levers and gearing. There was also the large inverted vessel into which the rising tide compressed air which was piped to suitable machinery.

Potential Tidal Power Sites
It is possible in theory to harness tidal energy almost anywhere. In practice the power that can be extracted from tidal action proves too expensive to compete with energy from more conventional sources unless the difference between high and low tide is considerable; engineers are generally agreed that the average tidal range must be at least 10 metres if there is to be a possibility of tidal energy conversion becoming an economic proposition.

Along a 1,600 km stretch of the north-west Australian coast, where a combination of geographical circumstances causes an average high- to low-water difference in sea level in excess of 10 metres the energy that could conveniently be converted into electric power has been estimated at nearly 100,000 MW — over five times the total consumption of Great Britain.

As is unfortunately the case with many of the world's best

tidal power sites, the Australian north-west coast borders an area of virtually no population and no electrical demand. It is a huge tract of semi-desert with Port Darwin, the nearest small city, some 800 km by land from the northern end of the strip of coast where conditions are most favourable for tidal power production. Perth is 1,500 km to the south west of its other extremity. If the capital could be raised to exploit the tidal energy going waste along this coast there would be enough to desalinate sufficient sea water to provide 75,000 sq. km of that potentially fertile semi-desert with 500 millimetres of irrigation water each year. Perhaps the recent discovery of extensive iron ore deposits in the area will one day create sufficient electrical demand to attract international finance.

To make tidal energy conversion economically justifiable, not only must the tidal range be high but there must also be substantial estuaries or inlets which can be conveniently dammed. Sites fulfilling both conditions are to be found on hardly 5 per cent of the world's coastlines. They are widely scattered and many of them, as in N.W. Australia, are far from centres of population, let alone industry.

In fact only four potential sites are near enough to major electricity demand areas to have occasioned serious thought of development. These are the Bay of Fundy (lying between Nova Scotia and New Brunswick, Canada, and extending south west to a little beyond the U.S. border at Passamaquoddy Bay); parts of the west coast of England and of the north-west coast of France; the land-locked White Sea, frozen over in winter, in the north-west corner of the Soviet Union; and the coast near Shanghai, China.

Modern Practice

To convert energy economically into electric power, you must have a large tidal basin; you must have engineering works to control the flow of a large volume of water in and out; and you must have turbo-generators designed to operate efficiently at a very low head of water. In the simplest form of tidal power station the water is allowed to flow into the basin naturally. At high water the gates are closed and when the sea level has dropped sufficiently to create a usable head

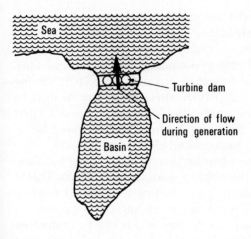

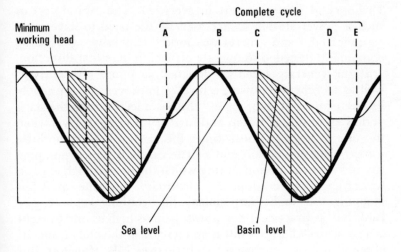

A-B Basin filling, sluices open
B-C Sluices shut, basin level held
*C-D Generators operating
D-E Sluices remain shut, basin level held

*About 35 per cent of total cycle

Fig 18. Simple one-direction tidal power plant

(3 metres is about the minimum) the trapped water is diverted into the turbine intake channel (Fig. 18).

There are two obvious disadvantages of this system of power generation. Not only is the power output intermittent, but the timing of this output is entirely dependent on the ebb and flow of the tide which alters from day to day. In practice this system will only provide power for about eight hours each day, and these eight hours will as likely be at times of minimum, as of maximum, electricity demand.

The eight useful hours can be increased by using the flow of the tide, as well as the ebb, to work the generators. This is done by closing the sluice gates at low tide and so holding down the basin level until sea level has risen enough to provide a working head. The energy which can be converted from inward flow proves in practice to be less than that available from outward flow; the best that can be achieved is 13 hours of power output in every 24. The capital cost of such a power station is increased by the need to have more complicated, and therefore more expensive generating equipment, designed to operate efficiently in either direction.

If the generators can be designed to pump efficiently as well as to generate power, a further improvement is possible by the process of 'topping up' the basin level during the slack period around high water, and further lowering its level when the tide is down. At both times the natural head is too little for efficient generation and a little carefully timed pumping can increase the period during which the head is high enough for efficient generation. As electricity must be used for pumping this looks suspiciously like robbing Peter to pay Paul, but in practice a net power gain as high as one to eight can be achieved. The result is an increase in the daily hours of efficient power generation to a little over 14½. However, this useful daily period will still vary its timing with the natural ebb and flow of the tide, and there will be periods every month when the generators can produce no power during the hours of maximum demand.

There is a fourth system of operation which can be regulated to provide a continuous output of power. This is the two-basin system (*see* Fig. 19).

In this case the rising tide is allowed into the first basin (W

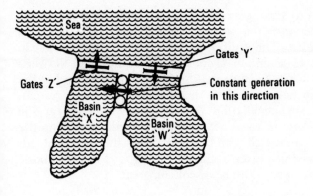

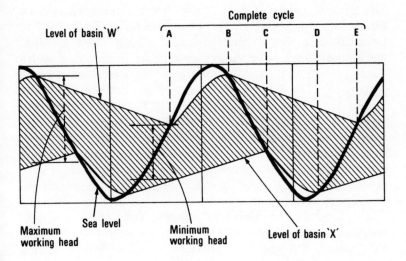

A-B Tide rising, gates Y open, basin W level rising
B-C Tide falling, gates Y closed
C-D Tide low, gates Z open, basin X level falling
D-E Tide rising, gates Y and Z closed

Fig 19. Twin basin tidal power plant with continuous power generation

in the diagram) through gates Y. From basin W it flows, via the generators into the second basin X, and from there via gates Z back into the sea on the falling tide. This system can be regulated to maintain a constant working head between basin W and basin X. However this head fluctuates and at the highest it can never equal the head in a single basin system. The twin basin installation consequently operates at the expense of a loss in total power output quite apart from the considerably increased capital cost. But the practical advantage of having power available throughout the 24 hours, despite the changing tides is, of course, great.

Contemporary projects

The earliest tidal-electric schemes were announced, almost simultaneously about 1920, in the U.S.A., in France and in Great Britain. The French envisaged a large plant at Aber-vrach, near Brest. The American scheme proposed a multi--basin system at Passamaquoddy Bay; the Canadians were to participate. The English planned a 13 km barrage across the mouth of the Severn estuary, which lies between England and South Wales, along with an associated 'pumped storage' reservoir in the nearby Welsh mountains. All three proposals were considered by the respective governments but were rejected on financial grounds. Thermal power stations with equivalent output were cheaper to install. There was no shortage of coal.

The English and American schemes were re-examined in the mid-thirties; this time they were pronounced economically sound. But again they were shelved. The Severn scheme was reconsidered once more in 1945, and once more pigeon-holed. In 1963 the Passamaquoddy project received the green light from President Kennedy, but the U.S. Congress could not find the time — or was it the inclination? — to vote the necessary funds.

Meanwhile the French authorities had shifted their sights from Aber-vrach to a more modest scheme on the La Rance estuary between St. Mâlo and Dinard on the north Brittany coast. Work on the dam and power house began in 1960 and the world's first commercial tidal power was fed into the French electricity grid in 1966. The French seemed pleased enough with the result for they soon announced a more ambitious project designed to impound 400 km² of sea in the Mont St. Michel Bay, east of La Rance. More recently, however, there have been reports of dissatisfaction with the efficiency of the La Rance power station.

The Soviet Union was not far behind the French. For in 1964 they installed a pilot tidal power plant at Kislaya Bay in the White Sea, south-east of Murmansk. While there have been no recent public reports of progress Russia announced an extremely ambitious tidal power project some years ago. The plan was to develop the great potential of the huge Mezen Bay on the eastern shore of the White Sea at the mouths of the Mezen and Kuloi rivers.

First there would be a 10 km dam across the Mezen inlet, with a planned tidal power generating capacity of 1300 MW. This was to be followed, once sufficient experience had been gained, by a second dam across the Kuloi inlet where a shorter dam but higher tides would produce a similar power output.

Finally the entire Mezen Bay would be enclosed by a 100 km dam forming a 2000 km² shallow basin. Two thousand low-head generating sets were to be distributed along this dam with a total power generating capacity, during maximum tidal flow, of 14,000 MW — equivalent to a 24-hour average output of about 4,000 MW.

The lack of news of this project may be due to unresolved problems caused by the extreme winter conditions. While tidal flow continues under the winter ice, and the river water should somewhat reduce the ice in the bay, much research was necessary on the behaviour of reinforced concrete covered with ice and subjected, below, to sea-water at sub-zero temperatures.

The La Rance Power Station

The La Rance river flows into the English Channel via a partly land-locked estuary which experiences an average tide range of 10.9 metres, the maximum spring tide range touching 13.5 metres with a water flow rising to about 18,000 m³/sec.

An important consideration in the financing of this power

station is its use to carry a major road across the estuary, shortening the St. Mâlo-Dinard run by about 30 km. The fuel saving to motorists and truck drivers represents a substantial part of the interest charges on the power station's capital cost (quite apart from time saving) thus 'justifying' the high cost per installed kilowatt capacity.

As tidal flow is a complex phenomenon depending not only on the gravitational pull of the sun and moon but also on the profile of the sea bed, on the increase caused by the water flowing into a decreasing 'funnel' between land masses, on the centrifugal force resulting from the rotation of the earth and on the natural resonance of the mass of sea in any semi-enclosed bay or inlet, work on the La Rance scheme began, wisely, with hydraulic model studies. The most important finding was that a barrage across the La Rance estuary would in no way disturb the rhythm of the local tides, a possibility that could have made the project impractical.

Research into low-head river generating plant had been carried out by the French over a long period of years, resulting, by 1957, in the evolution of a prototype generator suitable for tidal work. Tests and further development produced the present design in which the generator is mounted horizontally in a 'bulb', pointing into the main water flow. The turbine, not unlike a ship's propeller, is a four-bladed variable pitch screw. The bulb is supported from above and is fitted with twelve adjustable vanes which can be rotated to deflect the water flow to the most effective angle for turbine operation. These vanes, rotated fully through 90°, shut off the water flow completely, thus acting also as sluice gates.

The 24 bulb units at La Rance are each rated at 10 MW output for a head exceeding 7 metres. They can operate down to a head of only 1 metre, but even at 3 metres the output drops almost to 3 MW. They operate in reverse, though at a slightly reduced efficiency, and are designed for pumping service at heads up to 6 metres.

Overall Design
The La Rance installation is made up of four elements. To the west is a navigation lock, intended chiefly for fishing and

tourist boats. East of the lock is the power station proper, with the 24 bulb units fitted in the base of a 370 metre long hollow dam. A generator access tunnel extends through the length of this structure, the bulb units being fitted in channels below. Between the east end of the power station structure and a rock — the Rocher de Chalibert — the 178 metre gap is closed by a concrete-core rock-filled embankment. Between the Chalibert Rock and the eastern shore is a 150 metre control barrage in which are six vertical sluice gates.

The concrete used on the seaward side of the La Rance power house has to withstand, not only the mass of sea at high water but also the corrosive action of pounding salt water waves. Corrosion resistance was necessarily a problem in the design of the submerged metal parts of the generating plant, too.

Extensive research was carried out to find the right solutions to both problems. A leaking dam would have been an extremely costly repair job, and the frequent replacement of turbine blades due to corrosion would have been an equally unjustifiable expense. French engineers found answers to each problem. For the concrete it was a question of choosing the right aggregate and sand, and of specifying the right water-cement ratio and mix proportions. As for the metallurgical problem, two alloys were found suitable for the turbine blades — an aluminium bronze, and a high chrome stainless steel with nickel and molybdenum. Both were used. The bulbs' steel casings are, of course, protected by a paint unaffected by salt water.

The La Rance power station was designed to give a maximum power output, when all 24 generators are operating, of 240 MW. In practice, allowing for the slack periods of changing tides, the estimated total annual output is about 540,000 MW/hours. La Rance has proved that the system works, and that it may well be worth duplicating in other places.

CHAPTER 8

Earth Heat and Solar Energy

A glance back at the diagram on page 43 (Fig. 4) will show why I have put energy from the sun into the same chapter as 'geothermal' energy — heat from the interior of the earth. Both are the result of natural nuclear reactions — fission in one case, fusion in the other.

We explained nuclear fusion and fission in the second half of Chapter 3. There we saw that in the sun it is the fusion of hydrogen nuclei which releases the vast quantities of energy that radiate from it in the form of electromagnetic waves. We also learnt how the heavy nuclei of radioactive elements like radium, thorium and uranium tend to divide naturally, forming the nuclei of lighter elements. In nature this reaction is slow and as natural radioactive elements are themselves present only in low concentrations, the energy produced by natural fission seems at first to be of little significance. I say 'at first', because a closer look at the known facts reveals quite a different picture.

Measurements of the temperature gradient in deep bore-holes made in the earth's crust show that heat is being conducted up continuously from the interior at the rate of about 100 kilowatts per square kilometre. Knowing the low average thermal conductivity of the earth's crust, this heat flow from inside the earth would only be possible if the temperature 100 kilometres down were about 2,000°C. If this were so the rock at this depth would be molten. Seismic tests show that the earth is solid to a depth of about 900 kilometres, inside which is a 'plastic' layer 2,000 kilometres thick, before truly molten material is reached.

In view of this anomaly scientists have come to the only possible conclusion — that most of the heat that reaches the earth's surface from below is generated in the upper layers of the crust, only a few kilometres down. This can only be explained by the presence of radioactive elements undergoing natural decay.

Geothermal Steam

What this earth heat can mean in practical terms is well demonstrated in Italy, which has no coal or oil of its own and which was therefore 'driven', as it were, to find a secure alternative energy source. Work on the exploitation of geo-thermal heat was started in Italy in 1904 when a power station was built at Lardarello in Tuscany. The alternators there are driven by conventional turbines powered by steam which emerges from specially drilled boreholes. The steam is formed by the seepage of underground water down through cracks in the deep rock until it meets hot material. Today the Lardarello plant has been expanded to produce an electric power output of about 500 MW; this is not a great deal by today's standards, but is enough to supply Italy's railway system with its needs (Fig. 20A).

Earth heat is also used for the generation of electricity in western U.S.A., (output reached 400 MW in 1973 and is increasing by about 100 MW each year), in New Zealand (160 MW) and to a lesser extent in Japan, Soviet Russia and Iceland. In Iceland it has also been used since the mid 1930s to provide residential heating; nine out of ten houses in the capital, Reykjavik, are warmed by a geothermal district heating scheme. Research in the field of earth heat is cur-rently being conducted in the area of the San Andreas geological 'fault' in Southern California. Experts there believe that natural steam tapped from the fault at this one location could provide enough free energy to generate between 4,000 and 5,000 MW of low cost electricity — equal to the output of four or five giant nuclear power stations, and considerably more than the 3,000 MW of geothermal energy being made use of throughout the world today. Moreover there would be no problem of carrying in fuel or of radioactive waste disposal. The same experts are of the opinion that with relatively little additional research it should be possible to drill successfully for geothermal energy in many areas of the world.

In the past the system of exploiting geothermal heat has been to make boreholes into rock fissures; the presence of which is known by natural geysers or hot springs. These fissures connect with deep aquifers (water-bearing strata)

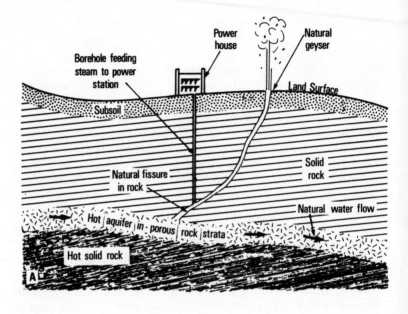

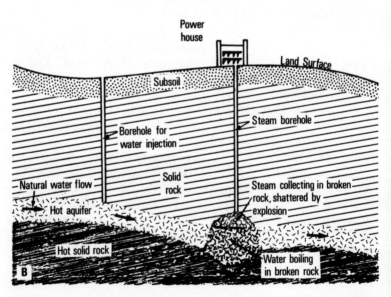

Fig 20. *Geothermal steam from (A) natural rock fissure (B) artificial 'boiler'.*

which produce steam because they are in contact with hot rock.

The latest development is to drill direct into a known hot water aquifer and detonate an explosion there to shatter the rock over a large area. Water from the aquifer can then seep much deeper into the hot rock below, producing more steam at a higher pressure. As the escape of steam lowers the aquifer pressure a second borehole is used to pump in fresh water. The result is a continuing supply of steam produced entirely by earth heat (Fig. 20B).

Solar Energy

In some of the newer houses in Japan the domestic hot water systems make use of the radiant energy of the sun. Rooftop solar panels have in them a zig-zag of copper piping filled with water, set in an insulated glass-fronted frame. The water, heated by sunshine, circulates to a fully insulated hot water storage tank in the house. So successful is the system that it is being extended for use on a larger scale, to provide hot water in blocks of flats.

Nor is this domestic use of solar energy confined to Japan.

A similar system has been used for years in Port of Spain, Trinidad, and doubtless elsewhere. An architect in London has been directing small scale solar panel research at the Polytechnic where he teaches. He reports that a solar-heated domestic water system in the English Midlands has been working well for over a dozen years, cutting the owner's heating bills to half what he would otherwise pay — and this is in an area where the average sunshine over the whole year is only four hours a day (Fig. 21)

In Washington, which has a little more sunshine than the British Isles, there is a privately built 'solar home' which has a hot water and central heating system on which the owner spent about $2,500 — just about twice the cost of a conventional system using oil as the energy source. The installation includes a normal central heating system but experience has shown that the fuel bill is only one third of what would ordinarily have been expected. This solar home is technically very simple. The pitched roof is covered with corrugated asbestos sheeting painted black, under glass. Water

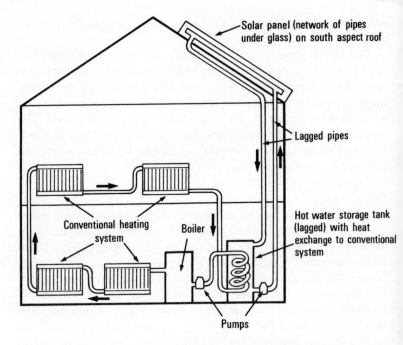

Fig 21. Principle of solar pre-heater for conventional hot water central heating system

is pumped up from a storage tank in the cellar and released so as to trickle down the grooves in the asbestos sheeting. The glass prevents evaporation but allows sunlight to warm the asbestos and so the water, which is then led back to the cellar storage tank. On most days the water temperature in the tank rises as high as 45°C.

The water used in the house's normal domestic hot water and heating system is pre-heated by passing it through a smaller tank submerged in the big cellar tank. All the boiler has to do is to raise its temperature that little bit more.

The sense of this use of free energy lies in the intensity of solar radiation. An average of nearly one kilowatt of heat falls on every square metre of the earth's surface whenever the sun is shining. (Due to the physical attitude of the earth in relation to the sun, the incident heat intensity is greater at the equator and less at the poles.)

The only .problem with solar energy is the problem of collecting it and converting it, at a competitive price, into a form we can conveniently use.

One line of research seeks ways of reproducing, on a mammoth scale, the schoolboy's method of making scraps of paper burst into flame by using a magnifying lens to focus on it a tiny spot of concentrated sunlight.

In an experimental solar 'furnace' set up in the French Pyrenees, the sunlight falling on a large hillside area can be focused, by means of mirrors, on to one small spot. Temperatures as high as 4,000°C are reported to have been achieved by this means.

In an attempt to design a more practical solar heat collector which could be used to produce steam to operate turbo-alternators or other machinery, research workers at a University in Arizona set up a system of adjustable mirrors. These concentrate the sunlight falling on the rooftop of one University block, into a double-skinned silvered tube, made like a huge thermos flask with an opening at the bottom as well as the top. The intense heat 'trapped' in this tube raises the temperature of pressurized gas circulated through a steel pipe inside it to levels which, in tests, have been as high as 1,000°C. The intensely hot gases are piped to a water tube furnace for the production of high pressure steam. By building large numbers of these solar heat collectors and dispersing them to cover, in all, the area enclosed by a 100 kilometre square, it is theoretically possible to produce sufficient steam to generate the United States' entire present demand for electricity, at a cost today estimated to be only 20 per cent higher than that of electricity produced by nuclear power stations.

Ocean Heat
Solar heat does not fall only on land and some scientists have not failed to appreciate the enormous quantity of heat energy which warms the waters of the oceans.

As recently as April 1974 an article in the *New York Times* gave the news that two American teams were working independently on projects to extract and use the heat of the Gulf Stream as it flows north east from the Caribbean.

The temperature of the surface water off Miami stays almost constant at about 25°C, while deep down the water flowing south to replace the Gulf Stream has a temperature averaging only 5°C. It has been calculated that the heat flowing northwards in the surface water between Florida and the Bahamas could be harnessed to produce all the electric power currently used in the United States of America.

The principle is quite simple. The warm surface water is to be used to 'boil' propane (or some other liquid that vaporizes at a sufficiently low temperature — ammonia for example). The vapour would be used, just as steam is used, to power specially designed turbo-alternators. The vapour would then be led back into condensers cooled by the deep sea water where it would condense and be ready for re-use.

The vast quantity of warm water which flows in the Gulf Stream and its relatively fast current would make pumping unnecessary, providing evaporating and condensing units were made large enough. One problem — the transfer of large quantities of energy from remote sites at sea to the mainland — could be solved, it is thought, by using the electricity produced to separate sea water into its constituent elements, hydrogen and oxygen, the gases being carried to shore by tanker for reconversion into electricity or for use as an independent energy source.

One of the American studies has calculated that the cost per kilowatt output of a Gulf Stream power plant of the kind envisaged would be only one third of that of a fossil fuel power station of the same capacity. Add to this the fact that the working energy source — Gulf Stream heat — would be entirely free, and one comes to the conclusion that power from this source could be significantly cheaper than any now produced.

Solar Electricity
Solar cells, which convert radiant energy directly into electrical energy are not new. The photographer's light meter has used the principle in a small way for many years. On a larger scale, communications satellites use solar energy to power their radio receivers, amplifiers and transmitters. Even the earliest active satellite, Telstar I, had 3,600 solar cells fitted

around nearly 90 per cent of its surface to provide electricity to operate the electronic equipment packed inside.

In theory the same system could be used in immense solar 'farms'. The energy would be free, but the capital cost would be immense. To meet the United States present total demand for electricity would need at least 10,000 square kilometres of solar cells — that is about the same area as for the solar heat collector system described earlier. In a country as large as the United States, with vast areas of desert, it would be feasible to disperse solar electricity farms in much smaller units over a wide area. Perhaps this will one day be done.

Another system which has been proposed by a commercial organization in the United States, is the production of electricity in enormous satellites, from which it could be beamed to earth in the form of high-powered electromagnetic radiation (Fig. 22). The proposed satellite would have to be 100 times as heavy as any yet in orbit, and would be assembled in space by sending up component parts on some 500 flights. This would be possible with the space shuttle planned to be working by the end of the 70s. The satellite would carry 64 square kilometres of solar cells very similar to those used to provide power for Skylab. Operating in continuous periods of sunlight, unimpeded by cloud of any kind, the very considerable electricity produced would

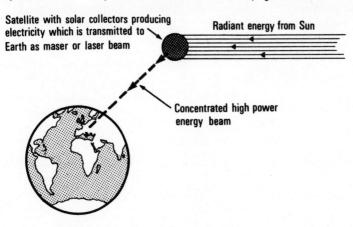

Fig 22. Principle of solar farm in space

be converted into microwave radiation and beamed to earth in so concentrated a form that it would carry many thousands of megawatts in a continuous power transmission. A power beam of this strength must not be allowed to stray away from the collectors set up on earth. To ensure this the satellite would be maintained in a high synchronous orbit, like the Intelsat communications satellites, which remain 'stationary' over a given point on earth because they are orbiting at precisely the right speed to keep up with the spinning of the earth. The direction of the microwave beam would then be controlled by a 'homing' radio signal from earth which, if it were ever lost by the satellite's control receiver, would immediately cause a fail-safe device to disperse the microwave beam over a wide area so that its concentration would no longer be dangerous on earth.

That a commercial firm is seriously planning such a solar power satellite which to us, today, sounds more like science fiction than a plausible possibility, demonstrates one thing: that free solar energy, which is available in such an unimaginably large quantity, and which is already today within technical reach at a price, will not remain unexploited for long. What is perhaps surprising is that so little serious research has yet been undertaken in this field — apart from research directed not towards solving the world's impending energy shortage, but towards enabling a handful of brave men to walk on the moon.

CHAPTER 9

Energy from the Atom

The sun releases energy by nuclear fusion. Radioactive elements in the earth's crust release it by nuclear fission. Was it unlikely, then, following the success of twentieth-century science in probing deeper and deeper into the secrets of the atom, that scientists should sooner or later find a way of harnessing one or other of the natural processes of fusion or fission for man's advantage?

That it should have been fission that was first exploited is not surprising. No heat is needed to split certain atoms; only neutrons travelling at a particular speed. Fusion only takes place at temperatures far higher than can be achieved by chemical combustion or even 'explosion'.

Decay and Fission

To understand the process that produces heat in the nuclear reactor, we must take a closer look at the materials that can be used.

The two most commonly occurring natural nuclear 'fuels' are uranium and thorium. Mixed in with natural uranium (which is mostly uranium-238) is a very small percentage of uranium-235. The essential physical difference between these substances can be seen from the following table:

Substance	Number of protons in nucleus	Number of neutrons in nucleus
Uranium 238	92	146
Uranium 235	92	143
Thorium 232	90	142

From their constitution you might expect them to be similar. In fact only uranium 235 is 'fissile'. It is the only one of the three that will split when hit by a neutron.

When a uranium-238 nucleus is hit by a neutron it absorbs it, turning into uranium-239 nucleus (92 protons, 147 neutrons). This is unstable and the odd neutron discards a 'beta' particle, which is identical to a free electron carrying its negative charge. When a neutron loses a beta particle it becomes a proton; so instead of 92 protons and 147 neutrons, we now have a nucleus made up of 93 protons and 146 neutrons. This is the nucleus of another element, neptunium, which is equally unstable. This nucleus discards a beta particle, so changing from 93 protons and 146 neutrons, to 94 protons and 145 neutrons. This is plutonium-239, a man-made element which like uranium-235 is fissile. When uranium-238 is used as a nuclear fuel, it is in fact the plutonium-239 formed which is split by neutrons. Uranium-238 is called a 'fertile' material to distinguish it from the fissile form, uranium-235.

(Thorium-232, the other naturally occurring fertile nuclear fuel goes through a similar chain of events when hit by neutrons. The neutrons are absorbed, forming thorium-233. This gives off beta particles, becoming, after two changes, uranium-233, another man-made fissile material.)

We saw in Chapter 3 that when the nucleus of an atom of natural uranium-235 (or for that matter any other fissile material, such as man-made plutonium-239, or uranium-233) is hit by a free neutron travelling at speed, the collision results in the formation of two smaller nuclei which share the original protons and neutrons. In fact there are always two or three neutrons left over and, when fission takes place, the enormous release of energy shoots these spare neutrons out at high speed. You would expect that the weight of the new nuclei formed, along with the weight of the spare neutrons, would add up exactly to the weight of the original nucleus. It turns out that this is not so; we find that the products of fission weigh slightly less, which means that some of the original matter has apparently disappeared. In fact it has been turned into energy, and it is this energy that appears in the form of intense heat and electromagnetic radiation when a fissile nucleus is split.

We started with a heavy nucleus being hit by a free neutron, and we ended with two lighter nuclei and two or three free neutrons. Suppose the split nucleus belonged to an atom in a lump of pure uranium metal. If one of the spare neutrons happens to hit another fissile nucleus in the metal we would expect this to be split in turn. This does not necessarily happen because to split a nucleus a neutron has to be moving at the right speed – called the critical speed. Fission does not take place if the neutron is travelling either too fast or too slow, and the neutrons shot out when a fissile nucleus is split are, in fact, travelling too fast.

Because of their high speed these spare neutrons will pass through the metal until they are either absorbed by non-fissile atoms or by impurities, or emerge from the metal's surface and are lost in whatever is around, or are slowed by collision with the atoms of the metal until their speed is reduced to the critical speed. Then they can split another fissile nucleus. In fact there has to be quite a thick lump of metal for the neutrons to be sufficiently slowed before they reach a surface and emerge, and this lump 'size' – the minimum size in which a spontaneous chain reaction can occur – is called the critical size (or mass) of the metal. Moreover it must be a lump of pure fissile material.

[In an atom bomb there are a number of small pieces of uranium-235 or plutonium-239, each too small to produce a spontaneous chain reaction. When the bomb is detonated some conventional explosive compacts all the small pieces of the fissile material into a single lump which is larger than the critical size. Then the atomic explosion takes place. Such an explosion is impossible in a reactor as the fuel is mostly fertile material in which the fissile material content never even approaches the purity required for an explosion.]

Slowing Down Neutrons
The aim of the nuclear reactor is, in fact, to establish a controlled chain reaction in low concentrations of fissile material.

This is done by dispersing the fuel in relatively small elements and surrounding these elements with a 'moderator' – a substance which slows down fast neutrons without absorbing them. While several materials can be used as a nuclear moderator, the two which have proved most suitable in practice are carbon and water. It takes just 180 millimetres of pure carbon (graphite blocks are normally used), or 50

millimetres of water, to reduce the speed of neutrons ejected
from a fission to the critical speed. Each of these moderators
has advantages. Pure carbon only slows down neutrons, and
does not absorb them at all; so by using it there is no neutron
wastage. The purity of the material is, however, important
for almost any impurities will absorb neutrons. Water, as we
have seen, reduces neutron speed much more quickly. This
means that a water-moderated reactor can be much smaller
than one in which carbon is used. Also water, even the purest
water, is much cheaper than carbon. On the other hand water
does absorb some neutrons, which means that to achieve a
chain reaction the fuel used in it must be more concentrated.
Natural uranium is mostly the isotope uranium-238 with only
about 0.7 per cent of fissile uranium-235. The natural
mineral will produce a chain reaction with a carbon moderator; but when water is used the fuel has to be 'enriched',
so that the content of uranium-235 is not less than 2 per
cent. Enrichment is a costly process.

While ordinary pure water absorbs neutrons, 'heavy' water,
in which the water molecules contain 'heavy' hydrogen
(deuterium) instead of ordinary hydrogen, does not. Heavy
water will therefore act as an excellent compact moderator in
which natural uranium can be used as fuel. But here, too,
there is a problem. Heavy water is only present as a very
small percentage of natural water, and the process of separ-
ating it is very expensive.

Controlling a Nuclear Chain Reaction
The practical layout of a nuclear reactor core is dictated
precisely by the moderator chosen. As it needs 180 mm of
pure carbon to reduce neutron velocity to the critical speed,
a carbon-moderated reactor core is built up by assembling a
huge mass of the material (built up from graphite blocks)
with 25 mm holes drilled through it at 205 mm centres.
Uranium rods (or tubes filled with uranium oxide) are then
pushed into the holes. As the uranium-235 present is a low
percentage of the whole an explosion cannot occur, though
the chain reaction will take place continuously, producing
energy, mostly in the form of heat, as it does so. When fast
neutrons pass through uranium-238, not all collisions result

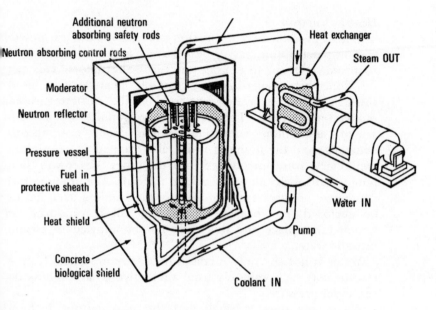

Fig 23. Essentials of nuclear power plant

in absorption. In fact many of the neutrons are slowed down; so uranium-238 acts as a moderator as well as absorbing some neutrons. We shall see later how this double action is put to use (For general layout of a power unit see Fig. 23).

Even with this arrangement there has to be some means of controlling the speed of the chain reaction. If it carries on too fast the whole reactor might grow too hot and melt, forming a dangerous radioactive molten mass. To control the speed of reaction we use rods of other metals which absorb neutrons, but which are not affected by them. The metals cadmium, boron and hafnium have this property. The control rods (only a few are needed) are fitted into extra holes in the graphite. When they are in position they absorb so many neutrons that the chain reaction cannot take place. By drawing the rods out gradually, a point is reached when neutron absorption is reduced to the extent needed for the chain reaction to start taking place. By drawing the rods out still further the reaction will speed up, producing more heat. In fact these control rods provide an extremely reliable and precise means of control.

Heat Extraction

The heat produced by the fission process in a nuclear power station reactor core has to be led away so that it can be used to produce steam to operate the turbo-alternators. This can be done by circulating any suitable heat-carrying fluid through the core. Water was used in the earliest carbon-moderated reactors. The holes in the graphite blocks contained aluminium tubes (aluminium does not absorb neutrons) and the uranium rods, (also 'canned' in aluminium containers) were introduced into the tubes. Water was then pumped through the tubes around the cans. The uranium rod containers are necessary not because the uranium itself has to be enclosed, but because radioactive ash is produced as fission takes place and this must not be allowed to escape into the water.

Water is not the only medium used to remove heat from a reactor core. A commonly used alternative is carbon dioxide gas under pressure.

Where the core is small and the heat output high, a medium is needed which can absorb and carry away heat faster than pressurized gas or water. As we shall see, liquid sodium metal is one answer.

The Carbon-moderated Reactor

The world's first reactor was not built to produce energy. Dr. Enrico Fermi, an Italian physicist, was working in the United States of America during the Second World War, when the U.S. Government were conducting research which led to the production of the first atom bomb.

Scientists already knew that uranium-235 was fissile. But its separation from natural uranium, of which it comprised less than 1 per cent, was a difficult and costly process. In 1940 a team of U.S. chemists, led by Glenn Seaborg and Edwin McMillan discovered plutonium, the man-made fissile substance which was produced when uranium-238 was bombarded by neutrons. Dr. Fermi was seeking a means of producing plutonium in quantity for the manufacture of atomic bombs when he designed and built the first reactor at Chicago, in 1942. Dr. Fermi's success is reflected in the fact that the first experimental atomic bomb, detonated at

Alamogordo, New Mexico, on July 16, 1945, and the third —
the bomb dropped on Nagasaki — was made with plutonium.
The second bomb, which wrecked Hiroshima on August 6,
three days before Nagasaki's destruction, contained uranium-
235.

Dr. Fermi's first successful reactor was carbon-moderated,
the rare metal cadmium being used to control the chain
reaction. The core was made from 100 tons of pure graphite
blocks with holes drilled in them so that rods of natural
uranium, when inserted, would be 180 mm apart. To achieve
a chain reaction six tons of the natural uranium had to be
inserted. But since there is only about 0.7 per cent of the
fissile isotope uranium-235 in the natural mineral, the heat
output of Dr. Fermi's first reactor was only 200 watts, so
there was no need for cooling. Once the principle had been
shown to work the United States built three large reactors for
the production of plutonium. These had a power output
measured in megawatts and so had to be cooled. Water was
used, pumped through aluminium tubes which lined the holes
in the graphite. Rods of natural uranium, also canned in
aluminium, were introduced into the water tubes.

Though the heat produced from these early reactors was
not wanted scientists at once realized their potential value as
an energy source. One gram of uranium-235, destroyed by
fission, produces as much heat as is released by about 3.6
metric tons of coal. It was not surprising that, immediately
after the war, scientists in the U.S.A., in England and in
Soviet Russia, all began active research on the production of
electricity from nuclear energy. Nor was it surprising that,
with research being conducted quite independently, the
practical outcome was three very different kinds of reactor.
In the event it was the Russians who commissioned the first
nuclear power station, in 1954, the same year that the
Americans launched their first nuclear powered submarine,
the *Nautilus*. Two years later, in 1956, the English commissioned the world's first large-scale commercial nuclear power
station.

Four Forms of Reactor
There are three basic essentials of any nuclear power plant.

1. There must be a reactor core to produce heat by controlled fission.

2. There must be a system to carry the heat out of the core so that it can be used to produce steam.

3. Finally, there must be a turbo-alternator to convert the energy of the steam into electrical energy.

We have seen already that there are two different kinds of naturally occurring uranium. One is fissile, the other non-fissile. Some reactors are designed to operate with natural uranium, containing only 0.7 per cent of the fissile isotope: others need enriched uranium which is more efficient but more expensive.

After the fuel the most important constituent of a reactor is the moderator. Here we have three practical alternatives. Pure carbon is one; pure water is another (water will normally be natural water, but 'heavy' water is an expensive alternative); the third, as hinted on page 99, is uranium-238.

Finally we have to provide a means for heat transfer. In this field we have three practical alternatives. Water is one, pressurized carbon dioxide gas another. The third, more expensive and technically more difficult to use, is the metal sodium in liquid form.

The problem of nuclear power station design therefore boils down to finding the most practical and most economic combination of the various alternatives. One of the costly considerations is that of safeguarding the public from radiation, and clearly the smaller the reactor, the cheaper will be the shielding.

There are, in fact, four different types of reactor in regular use today for the production of electricity. The practical combinations of the possible alternatives are shown in the following table:

	Original Design	Fuel	Moderator	Heat transfer
A	British	Natural	Carbon	Carbon dioxide
B	American	Enriched	Water	Water
C	Russian	Enriched	Carbon	Water

A fourth type, now under active development in several countries, uses enriched fuel, moderation being effected by fertile uranium and heat transfer by liquid sodium. This is the 'fast-breeder', the principle of which is described below.

The Carbon-moderated Gas-Cooled Reactor

I have already described the practical design details of a carbon moderated reactor core. The first commercial nuclear power station having reactors of this kind and providing a steady power output of 70 MW of electricity from two reactors, was commissioned at Calder Hall in north-west England in 1956. Two more reactors were added later. Each core, built up of 1,000 tons of graphite blocks, is in the form of a huge vertical cylinder, 8 metres high and 11 metres in diameter. In this are 1,696 vertical fuel channels, each accommodating six rods of natural uranium in cans of a magnesium-aluminium alloy each about one metre long. In all, the core holds about 130 tons of uranium. The carbon-dioxide gas used to cool the core is pumped in below at a pressure of about 3.5 kg/m^2 (roughly 6½ atmospheres) and at a temperature of approximately $100°$C. Passing up through the gaps between the fuel cans and the holes in the graphite it becomes heated to about $330°$C. It is then led to a heat exchanger where it passes over steel water pipes, boiling the water and producing high-pressure steam at about $310°$C and at a pressure of about 7 kg/m^2.

Around the core and the gas inlet and exhaust chambers is a 150 mm thick steel pressure vessel a little over 11 metres in diameter and 22 metres high. Outside this is a heavy solid concrete container called a 'biological shield' designed to prevent the escape of any radiation which may have got through the walls of the pressure vessel. Control of each reactor is by means of fifty boron-steel control rods, which can be easily lowered into additional holes in the graphite, or withdrawn, from above, through corresponding holes in the 'roof' of the concrete shield. Gear for putting in fuel rods, or for removing them when spent, is also fitted above.

The Water-moderated Water-cooled Reactor

Though the United States had acquired considerable know-how in the field of carbon-moderated reactors during the war, their first objective, in designing a reactor for energy, as opposed to plutonium production, was to power a submarine. As such a power plant had to be as small and as light as possible they inevitably decided to use water as the moderator. When the *Nautilus* was launched in 1954, the new reactor proved so successful that American engineers decided to develop a similar, enlarged, design for use in electric power stations.

The first U.S. nuclear power station, at Shippingport on the Ohio River in Pennsylvania, was commissioned in 1957. Since only 50 mm of water are needed to moderate fission neutrons to the critical speed, nine fuel rods can be accommodated in a water-moderated core in the same area as occupied by only one in the carbon-moderated type. Con-

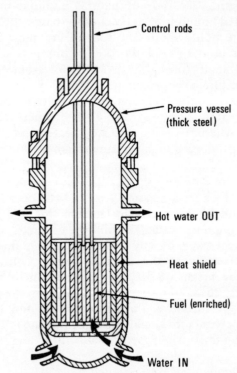

Fig 24. Pressurized 'light' water reactor

sequently, though the heat output of the Shippingport reactor was higher than that of each of those at Calder Hall, the pressure vessel built to hold the core was only 2.5 metres in diameter and 9 metres high. The water used as moderator was also used as the heat transfer medium, and so was pumped continually through the core and piped to a heat exchanger. The circulating water was allowed to heat to 330°C as it passed through the core, and to prevent it from boiling, it was pressurized to about 70 kg/m^2. This is a very high pressure, and to ensure safety the pressure vessel was made of steel 200 mm thick. It has a domed head, which is bolted on· so that it can be removed for the replacement of the fuel, and has glands at the top through which control rods pass (Fig. 24).

As water absorbs neutrons to some extent, the fuel chosen for use was the expensive enriched uranium, and as uranium metal will react chemically with super-heated water, uranium oxide was used instead. This extremely stable compound was canned in tubes made of an equally inert alloy called zirconium.

Because of the extremely high pressure inside the early U.S. reactors, a new design was introduced in which the pressure was only 35 kg/m^2 and in which, consequently, the water began to boil near the top of the core as it circulated upwards. While water was pumped in at the bottom, steam flowed out at the top, this being condensed when it passed through the heat exchanger.

The Carbon-moderated Water-cooled Reactor

In Russia engineers had worked on the original American concept of a carbon-moderated water-cooled reactor. In those first American reactors, designed for the production of plutonium, the heat was not wanted. Water was pumped through so fast that it did not have to be pressurized to prevent it from boiling. The Russians, to avoid the need to build an enormous pressure vessel around a large carbon-moderated core, as at Calder Hall, made their fuel elements in the form of slim cylinders. These were located inside stainless steel tubes through which the pressurized cooling water was circulated down the outside and then up through the middle of the fuel. Though steel absorbs neutrons, the tubes were not so thick as to prevent a chain reaction. But the fuel, as in the water-moderated reactor, had to be enriched so that the percentage of uranium-235 was higher than it is in natural uranium.

In the first Russian power reactor, which had a relatively small output of 5 megawatts, the cooling water entered the core at a pressure of about 50 kg/m^2, and a temperature of about 115°C, leaving at 260°C. In a later design the steam produced in the external heat exchangers was piped back into the reactor, again through stainless steel tubing, and super-heated to 500°C at 45 kg/m^2 pressure, before being used to power the turbines. The first reactor of this type was commissioned in 1958 and had an output of 100 MW. Subsequently Russia built pressurized water reactors of twice this output using a design similar to that now common in the United States.

Each of the three original reactor designs had its advantages (A) and disadvantages (D), which can be summarized as follows:

	English	American	Russian
Moderator	Carbon (Large core) D	Water (Small core) A	Carbon (Large core) D
Coolant	Gas (Large medium pressure vessel) D	Water (Small high pressure vessel) D	Water (Confined to pipes) A
Fuel	Natural A uranium	Enriched D uranium	Enriched D uranium
Biological Shield	Large D	Small A	Large D

The Fast Breeder Reactor

All three basic designs have been modified and improved, but the most significant new design has been that of the fast breeder reactor, the first of which was successfully commissioned in Scotland by British engineers. The fast breeder is designed not only to generate heat from nuclear fission, (which is the aim of all the earlier post-war reactors) but also to produce fissile plutonium-239 from fertile uranium-238 (as in the war-time American reactors).

In a conventional reactor about 40 per cent of the neutrons released by fission hit fissile nuclei after their speed has been moderated, to cause further fission and so keep the chain reaction going. A further 28 per cent of the released neutrons are absorbed by fertile nuclei which decay to form man-made fissile material. The remaining 32 per cent are lost. They may be captured either by 'foreign' nuclei (for example, those of the control rods and containment vessel), or by fissile nuclei which they have hit at the wrong speed; or they may escape from the core altogether. In the fast-breeder the core of highly enriched material, which is extremely compact, is surrounded by a 'blanket' of fertile material which captures a large proportion of the neutrons which would otherwise be wasted. So compact is the core that its heat has to be removed very rapidly, and it is for this reason that liquid sodium metal is used as the heat carrying medium — sodium being the only metal which melts at a low temperature and which also does not absorb neutrons greatly.

So efficient is the fast-breeder that its production of fissile material (plutonium-239) is greater than its consumption of fissile material. As a result, though it actually 'burns' fissile material to produce heat, it provides this itself (with some to spare as a by-product) and need only be supplied with the more readily obtainable uranium-238.

While the fast breeder is still in in the development stage, it promises to be the power reactor of the future. There are new problems in using liquid sodium as the cooling medium but there is also an advantage in that the fast-breeder core is small (no larger than a domestic dustbin) and needs no pressure vessel, since the liquid sodium is circulated at low pressure. In fact the entire reactor is so compact that, complete with a concrete biological shield, the early experimental model built in Scotland was contained (for safety) in a 12 mm thick steel sphere only 55 metres in diameter.

This steel container and the biological shield were together very much less costly than either the heavy steel pressure vessels used for pressurized water reactors, or the huge biological shield needed for carbon-moderated reactors.

Over the years power reactor designs have been continually improved and tested and there is little doubt that not only can nuclear power take over from thermal power, as fossil fuels become exhausted, but it is also capable of supplying all man's energy needs in the foreseeable future. There is, however, one considerable problem that follows from such a substitution.

Nuclear reactor fuel does not 'vanish' as it 'burns'. Even after chemical removal of all remaining uranium (or plutonium) from spent fuel rods, a considerable quantity of highly radioactive 'ash' is left, and this has to be disposed of safely. The uranium (or plutonium) in this 'ash' is first removed by acid treatment, and the resulting 'slurry', which will remain dangerously radioactive for up to 100 years, is usually pumped into deep underground steel and concrete tanks, or into drums which can be dumped far out at sea. As drums in the sea may become corroded and eventually leak, only 'low-level' radioactive waste is disposed of in this way, the more dangerous material being kept underground.

This question of disposal presents a serious problem, for by the year 2,000 A.D. nuclear power stations around the world are expected to be producing 200,000 cubic metres of radioactive waste every year. Indeed this is the biggest single argument in favour of trying to develop the production of electricity from the 'clean' hydrogen fusion reaction, or alternatively from the radiant energy of sunlight.

CHAPTER 10

Looking Ahead

That the 1974 world energy 'crisis' is a temporary problem is perfectly clear. The United States is capable of replacing all oil previously imported from the Middle East by oil recovered from the enormous oil shale deposits she possesses. Canada with her tar sands is in a similar position. Even Britain, with her newly found undersea oil fields, is capable of producing a substantial proportion of her needs within five years. Import of the remainder should not prove too difficult and nuclear power will more than replace electricity now produced by burning oil.

Yet even the new British offshore oil fields; even the American oil shales and Canadian tar sands; even the world's very considerable reserves of coal will one day come to an end. What will happen then?

We have already seen some of the possible solutions to this problem.

Control of wastage could provide one significant contribution. An immense amount of heat is wasted as a result of poor domestic insulation alone. Some experts believe that half the heat we put into our houses today could quite easily be saved. One does not have to throw heat away in order to enjoy a comfortable temperature. Coffee in a vacuum flask stays hot without putting it on the cooker.

Energy recycling can play an important part. In Paris all the city's refuse is dumped at a huge incineration plant not far away. The incinerators are more or less conventional; only the heat produced is not wasted. Instead it is used to produce steam to drive turbo-alternators which supply about 10 per cent of the needs of the metropolis. One sewage farm in London processes the effluent of three London boroughs to produce 180,000 cubic metres of gas every day. At most sewage farms around the world the gas produced bubbles quietly into the air where it is lost.

At many thermal power stations there are huge concrete cooling towers designed to disperse excess furnace heat into the atmosphere. At other power stations river water is pumped through instead, the waste heat flowing into the river. At Battersea power station on the Thames in London there are no cooling towers, and the water pumps do not circulate river water. Some twenty years ago plant was installed to pump water heated by the power station's spent furnace gases to the buildings of nearby Pimlico. There it is distributed to many of the buildings in the area to provide the heat to operate conventional heating systems. There are thousands of power stations around the world where similar 'district heating' schemes could profitably be installed.

We know that there is still considerable scope for the establishment of hydro-electric power stations in many countries of the world. French experience in low-head bulb generators, suitable for use in major rivers without the need for expensive high-head dams, is of considerable significance here. There is enough low-head water flowing steadily down the river Amazon to provide, at a fraction of the present cost, one third of the electrical energy currently consumed in the United States of America.

We have seen how the prevailing winds that blow reasonably steadily in many parts of the world could be put to use to provide more 'free' electricity. And we have examined the considerable potential that exists for electricity production from the ebb and flow of the tides.

We have looked at geothermal energy, and have seen how near technology is to a breakthrough which would make it possible, in many parts of the world, for us easily to tap the earth's natural heat to produce enough steam to run turbo-alternators. We have seen how the heat of the Gulf Stream could be extracted for our use.

However none of these means has the potential to provide man with more than a part of his total energy needs. To meet all his foreseeable demand, once the fossil fuels are exhausted, there are only two sources. These are the sun and man-made nuclear energy.

We have seen many problems that need to be solved before we can make use of solar energy on a large scale. Doubtless they will be solved. Equally certain is it that much more use will be made of the sun's energy on the personal scale — by using solar heating and solar electric panels for the supply of energy to individual homes.

Nuclear power stations are with us already and development is continuing apace. With the emergence of the fast breeder reactor, for which the necessary fertile material is readily available in quantity, and which itself produces all the fissile material needed in its core, the question mark that hung originally over the supply of sufficient nuclear fuel for the fission reactor is vanishing far into the future. Only the problem of disposing of steadily increasing quantities of radioactive waste remains to be solved. Could we, perhaps, send up huge rocket-loads into the sun? Already we possess the technology to do this.

With water so plentiful on our planet the nuclear fusion reaction would seem to offer us a virtually unlimited energy source. But science has not solved the immense technical problems of sustaining, controlling and containing the hydrogen fusion reaction which requires a temperature of several million degrees centigrade. A great deal of money and research is now being poured into this problem in the United States, in England and in Russia, the idea being to contain the reaction in mid-air in what is sometimes called an electro-magnetic 'bottle'. So far the research has hardly begun to solve some of the more formidable problems and Herman Kahn, director of a well-known American technical 'think-tank', writing not very long ago, did not list the hydrogen fusion power station in his one hundred technical innovations very likely before the year 2,000. He did, however, include it in the list of 25 less likely, but possible achievements.

In Chapter 4 we spoke briefly of the problem of replacing oil as the principal energy source for small-scale transport. We came to the conclusion that in the absence of a practical small nuclear power plant the alternative lay in storing energy either chemically in a man-made fuel such as hydrogen, or in some form of electric battery.

The concept of the battery-powered electric car has many attractive features. An electric motor is far less complicated than a petrol engine and needs less maintenance. Power output covers a much wider range so that an electric car can do without gears. Even conventional friction brakes are less important because when an electric motor is connected to a battery with the positive and negative terminals interchanged the result is powerful smooth braking which, in fact, serves to charge the battery to some extent. There are no exhaust products from the electric motor, so it eliminates the pollution problem which has become very real now that there are so many petrol- and diesel-powered cars on the road.

In theory the lead-acid battery can be used as an energy source for an electric car. To be practical, however, an electric car must be able to compete on equal terms with a petrol-engined car. This means that it must be capable of travelling as fast, and must be able to travel as far on one charge as a conventional car will travel on one full petrol tank.

The maximum energy density of the lead-acid battery is about 20 watt-hours per kilogram of battery. The weight of the batteries a car can carry is limited because, to compete with the conventional car the weight of the electric motor and the batteries must together not be more than the weight of a petrol engine and its fuel. When this limitation is put into practice we find that an electric car capable of 125 kilometres per hour will have a range (with a full battery charge) of only about 60 kilometres on the open road or of a little over 30 kilometres in the start-stop situation of driving in city traffic. This is simply not good enough because the comparable petrol-engined car has a range of at least 300 kilometres.

There are, of course, alternatives to the lead-acid battery. The relatively new chargeable nickel-cadmium batteries (used in so-called 'cordless' appliances like grass-trimmers, hedge-cutters and razors, and in the photographer's electronic flash gun) are better, if more costly. But even with these the practical open road range of an electric car can only be increased to about 120 kilometres.

The silver-zinc chargeable batteries developed in the United States for space vehicles are one step better still, but these are unduly expensive and the practical range would be increased only to 160 kilometres on the open road, 80 in traffic — still not enough.

Under development there is an even newer re-chargeable battery — the zinc-air battery. This is capable of holding an energy density six times that of the lead-acid battery, giving an open road range of about 350 kilometres. If those working on the development of this battery are as successful as they hope, this may supply the answer we are looking for. Instead of selling petrol, garages on the road would exchange batteries. Spent batteries would be charged from the normal electricity supply which, once the potential of new forms of electric power production have been fully developed, would not be at risk due to diminishing reserves of the fossil fuels.

Fuel Cells

An alternative to the chargeable battery is the fuel cell which generates electricity directly by a non-combustion reaction between an oxidizable gas and oxygen, the latter obtained from the air. It seems likely that when this new form of energy converter has been fully developed it will be capable of an energy output of at least 160 watt-hours per kilogram, giving an open road electric car range of as much as 500 kilometres.

The fuel cell is not a 'battery' because you cannot take energy out of it without providing a continual supply of gas, and simultaneously removing the product of its oxidation. Nor can you 'charge' it without returning to it the original oxidation product and removing the gas produced. In other words, it may be thought of as a 'cold' process of combustion. When you burn hydrogen you must supply the gas, and also oxygen, slowly but steadily; the combustion product is water. The reverse process is the decomposition of water to produce hydrogen and oxygen.

One of the simplest forms of fuel cell is the hydrogen-oxygen cell. The gases hydrogen and oxygen (ordinary air is in fact used) are piped to porous electrodes in an 'electrolyte' of potassium hydroxide. The result is an electric potential across the electrodes. When current is drawn gas is slowly consumed and water is released in the electrolyte which becomes diluted. Provided the gases are constantly supplied and the water is removed from the electrolyte, the fuel cell will continue to produce electrical energy indefinitely (Fig. 25).

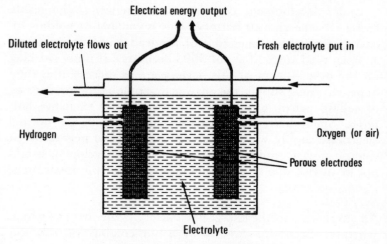

Fig 25. *Hydrogen-oxygen fuel cell*

There are practical fuel cells of various kinds, some of them using fuels which are considerably cheaper than hydrogen. Natural gas, volatile alcohols and propane have been tried. Some fuel cells use the reaction between a metal and a gas for the direct production of electricity (e.g. lithium and hydrogen, or cadmium and iodine). In cells of this kind the reaction product can be directly decomposed by heat to produce fresh fuel. Such cells can therefore be used to convert heat into electricity without an intermediate mechanical stage.

One great virtue of the fuel cell is its high energy conversion efficiency. The conversion of chemical energy into mechanical energy by the petrol engine involves an energy loss of 75 per cent. Even the more efficient diesel engine is only 35 per cent efficient, some 65 per cent of the oil's chemical energy being wasted. Most fuel cells, on the other hand, are not less than 75 per cent efficient and the efficiency of some is greater than 80 per cent. This efficiency to some extent offsets the high cost of the fuels. Even so hydrogen is at present so expensive (for a motor car it would be carried, of course, in liquid form) that the cost per mile in the case of a motor car powered by hydrogen-oxygen fuel cells was estimated, before the 1973-74 oil price rises, to be about ten times that of a petrol-engined car. At the same time the cost of fuel for natural gas-air fuel cells was estimated to be only one quarter of the cost of petrol for an equivalent power unit — also before the oil price rises. Since natural gas reserves are likely to be exhausted no less quickly than the world's oil, we are forced back, in the long term, to hydrogen. It is, after all, the world's ninth most plentiful element, freely available, at a cost, from water.

One very great advantage of the hydrogen-air fuel cell is the complete lack of pollution in its action. There is none of the poisonous fumes which the petrol engine produces: only pure water. Which is why it has proved so valuable in the U.S. space programme, producing fresh drinking water as well as energy for astronauts in space.

Though the hydrogen-air fuel cell produces a working potential of only about one volt, and has a correspondingly low power output, it can, of course, be used in series to produce a higher voltage output, and in parallel to produce higher power. So much so that it is considered perfectly feasible to use this fuel cell, in huge banks, to provide power stations with off-peak storage comparable to the method of pumped storage described in Chapter 6. The fuel cells would have the excess base load electrical energy fed into them at night, the hydrogen produced being led into huge underground gasometers. During peak load periods next day the hydrogen would be fed back, along with air, into the fuel cells to generate additional electricity.

The fuel cell's efficiency in converting chemical into electrical energy without the losses incurred from first producing heat and, from it, mechanical energy, is not matched by any form of solar converter yet developed. In fact typical experimental efficiencies so far achieved in solar generators vary from as low as 0.5 per cent up to 3.5 per cent at the most. As solar radiation is free, this inefficiency is not a serious drawback; it only means that solar converters must be built very much larger than would otherwise be necessary. But it is the reason why solar converters cannot be used to power mobile transport. The solar cells used on communications satellites produce enough electrical energy to power radio receivers and transmitters; but this is a fraction of the power needed to propel a vehicle.

MHD

There is, however, another method of generating electricity by direct conversion from heat, which may well be destined to play an important part in central power generation, though not for mobile use. It is called magnetohydrodynamic (MHD for short) generation.

The reader will remember that the principle of the electrical generator is the movement of an electrical conductor through a magnetic field. In the modern alternator it is usually the field that is made to move, the conductor being the wire in the coils, which are in fact stationary.

Metal is not the only conductor of electricity. Ionized liquids and gases can be excellent conductors and it is this fact that is made use of in the MHD generator. Air is heated in a furnace and then pumped under pressure through an expansion duct, rather like a huge rectangular horn. A powerful magnetic field is produced across this duct, and electrodes are located in the air flow at the top and bottom of the horn. Because hot gas is ionized and conducts electricity, the effect of this arrangement is to provide between the electrodes a conductor (the flowing ionized gas) moving rapidly through a magnetic field. The result is the

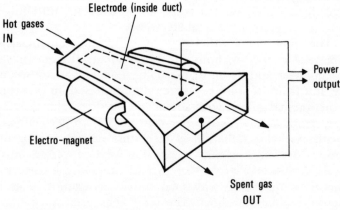

Fig 26. Principle of MHD generator

production of an electrical potential across the electrodes.

In a typical experimental MHD generator the air is heated to 2,000°C and 'seeded' with fine particles of potassium (which is subsequently recovered) to improve conductivity. About 60 kg of the seeded hot air is pumped each second through an expansion duct 3 metres long, the inlet being about 0.1 m² in area, the outlet about 0.3 m². With a powerful electro-magnet producing a field across this duct, the electrodes deliver 32 MW of electrical energy, of which one quarter is used to power the electromagnet, leaving about 24 MW as net output. To produce this amount of electricity by conventional means would need a substantial steam turbine and an equally substantial alternator. The MHD generator is smaller, simpler, lighter and has no moving parts. That this type of generator will soon be developed to the extent of producing as much as 1,000 MW from a single unit is now considered likely by experts. While the temperature of the exhaust gas remains relatively high, which means that the generator wastes much heat, it has many advantages. Moreover some of the exhaust gas heat can be usefully used to preheat the furnace intake air, or even for the production of steam.

Energy in 2025 AD

Any reader who has read this far should now be equipped to make as intelligent a guess as I can about the state of the world's energy situation fifty years from today. Some things are clear. Very few worthwhile reserves of oil or natural gas will be left. There should still be reserves of coal, but plans will be well advanced for running down coal production when ultimately it is finished.

Electricity will be produced in substantially greater quantity than today. While some coal-fired power stations may still be in operation, most will have been converted to use geothermal steam, and the nuclear power station will have become paramount. It will not, however, be based so much on the nuclear fission, as on the much 'cleaner' fusion reaction. It will use hydrogen, obtained from water, as the basic fuel, and the MHD generator will replace the turbo-alternator.

Electricity will also be produced on a large scale at solar farms, and on a smaller scale from hydro-electric and wind-power stations and, possibly, from a few major tidal power stations, and from one or more huge complexes designed to extract heat from the Gulf Stream.

Transport will be powered by a variety of means, the most likely being the small nuclear-electric generator for ships and possibly for large locomotives and aircraft. Jet engines will be of entirely new design, more akin to the hydrogen and experimental nuclear powered rockets of today, than to the kerosene fuelled propulsion units we now know.

All smaller means of transport, right down to the motor car, will also be powered by the electric motor; only in this case the energy supply is more likely to come from new forms of electrical storage battery or, equally probable, from various kinds of fuel cell. Hydrogen, piped through gas mains, will be widely used as a domestic and industrial energy source.

All in all, it seems there will be two principal basic energy sources: the oceans and the sun. The oceans will produce uranium (if this is still needed and man has found a simple answer to the problem of disposing of radioactive waste) and hydrogen (for both fusion reactors on the large scale and for the fuel cell on the small.) The sun, for its part, will continue to provide mankind, if he himself survives, with unlimited free radiant energy, for many millions of years.

So next time you go down to the seaside to swim and to bask in the sun, remember that there above you and before you lie sources of all the energy man will ever need.

INDEX